# DINOSAURS OF ITALY

**LIFE OF THE PAST**

James O. Farlow, editor

# DINOSAURS OF ITALY

Cristiano Dal Sasso

With the collaboration of
Giuseppe Brillante

Translated by Barbara Ferrett Rogers

INDIANA UNIVERSITY PRESS
*Bloomington and Indianapolis*

This book is a publication of

Indiana University Press
601 North Morton Street
Bloomington, IN 47404-3797 USA

http://iupress.indiana.edu

*Telephone orders* 800-842-6796
*Fax orders* 812-855-7931
*Orders by e-mail* iuporder@indiana.edu



The paper used in this publication meets the minimum requirements of American National Standard for Information Sciences—Permanence of Paper for Printed Library Materials, ANSI Z39.48-1984.

Manufactured in the United States of America

Library of Congress Cataloging-in-Publication Data
Dal Sasso, Cristiano.
[Dinosauri italiani English]
Dinosaurs of Italy / Cristiano Dal Sasso.
p. cm.
Includes bibliographical references and index.
ISBN 0-253-34514-6 (cloth : alk. paper)
1. Dinosaurs—Italy. I. Brillante, Giuseppe. II. Title.
QE861.9.I8D3513 2005
567.9'0945—dc22
2004014815
1 2 3 4 5 10 09 08 07 06 05

# CONTENTS

## Contents

# FOREWORD

DINOSAURS HOLD A certain fascination for almost everyone who knows of their existence. This is particularly true for children, some of whom, like me, go through life without losing that interest. *Tyrannosaurus rex, Allosaurus, "Brontosaurus," Stegosaurus,* and many more of their names are better known internationally than the capital cities of more than half the countries in the world (including Canada). Until recently, the best-known dinosaur names were mostly North American forms. This is a matter of historical development more than anything else. Although the first dinosaurs recognized were collected and scientifically described in Europe early in the nineteenth century, the specimens were incomplete and poorly understood. On the other hand, some of the first North American dinosaurs were almost complete skeletons that were well described and illustrated in scientific papers and were seen by millions of people in spectacular dinosaur displays such as those of the American Museum of Natural History in New York. Availability of research funds, Hollywood movies, a publishing world centered in New York City, and many other factors were also involved.

Like everything else, however, this situation has changed. Children are as challenged today in spelling and pronouncing dinosaur names and still take great delight in practicing these tongue-twisting names on their befuddled parents. But with new tools and influences, such as the World Wide Web and movies such as *Jurassic Park,* a wealth of new information is being disseminated at a greatly accelerated rate, mostly to the younger generations. *Tyrannosaurus* is still the most famous dinosaur, but *Velociraptor* has also become well known. And in my job as the Curator of Dinosaurs at the Royal Tyrrell Museum, I even frequently get asked about new dinosaurs such as *Scipionyx* that children learn about on the Internet.

We have in recent years entered a golden age in the collection and description of new types of dinosaurs, most of which are being found outside of North America. Some of these are being discovered in areas where we would not have expected to find them—such as the arctic islands of Canada or the ice-locked mountains of Antarctica. Other good specimens are being recovered from Europe, the region where the first dinosaurs were found. Fossilization is a process that relies on the presence of certain circumstances. In Dinosaur Provincial Park, the area of Canada where I do most of my fieldwork, the climate and environment were perfect to support dinosaurs 75 million years ago. The rivers of that time were full of sand and mud that buried the bodies of the dinosaurs after they died. And now those long-buried sands and muds have been brought to the surface of the earth again by erosion and are exposed in the badlands, where we can find the fossilized bones they contain. If all those conditions had not been just right, we would never have been able to recover the five hundred skeletons that we have from the Park. Yet the right conditions can sometimes occur in other ways. For example, *Scipionyx* is a tiny animal whose body was washed out to sea, far from the environment where it had lived. Nobody would have predicted that the specimen would be discovered in those marine rocks. And it was not found because someone was specifically looking for a dinosaur. Like many of the great finds, this one was made inadvertently by an amateur. The discovery of the first good Italian dinosaur, therefore, was made against the odds. But now that it is known that significant dinosaurs can be found in Italy, more people will have their eyes open looking for them, which will greatly increase the chance that more specimens will be found.

Fossil sites do not have to be rich to be important. Fragments of even a single bone from some parts of the world are enough to tell us that a particular dinosaur actually lived there. This can be important in determining the patterns of migration and dispersal. Southern Europe, for example, was a crossroads for dinosaurs moving between Africa, North America, and Asia. Understanding the movements of dinosaurs helps us determine where they originated and how they diversified. And because there is so much interest in the biology of these animals, whenever a specimen such as *Scipionyx* appears with parts of its soft tissue preserved, it will be the focus of international attention. Better-preserved specimens of certain kinds of dinosaurs, such as the feathered dinosaurs I work on from China and the closely related *Scipionyx,* can help us determine the relationships of dinosaurs to their modern descendants, the birds. Through these and other studies, we know that dinosaurs did not die out 65 million years ago but live on in their descendants, which include more than ten thousand species of living birds.

There are still so many things that we do not know about dinosaurs. It is a vibrant and growing field of study that opens the doors to forgotten worlds that were responsible for shaping the present. After all, do you really think you would be reading this book if *Tyrannosaurus* and *Velociraptor* had not died out so many millions of years ago?

Philip J. Currie
Curator of Dinosaurs
Royal Tyrrell Museum
Drumheller, Alberta, Canada

# ACKNOWLEDGMENTS

GATHERING ALL THE information contained in this book would have been an impossible feat without the help of all the people whom I should like to thank here, apologizing if I have overlooked anybody.

My thanks to the paleontologists and geologists who provided information and advice: Ronan Allain, Eric Buffetaut, Dan Chure, Phil Currie, Zhiming Dong, Alan Gishlick, Thomas Holtz Jr., Jahn Hornung, Eva Koppelhus, Walter Landini, Michele Lanziger, Martin Lockley, Alberto Lualdi, Dave Martill, Octavio Matéus, Christian Meyer, Ryosuke Motani, Stefania Nosotti, Anna Paganoni, Gregory Paul, Isabella Premoli Silva, Angelmario Rebecchi, Silvio Renesto, Olivier Rieppel, Martin Sander, Mary Schweitzer, Marco Signore, Rupert Wild, and Larry Witmer.

My special thanks go the scientific revisers who carefully read and commented the text: Marco Avanzini, Fabio Dalla Vecchia, Giuseppe Leonardi, Umberto Nicosia, Giorgio Teruzzi, and in particular Matt Carrano, who patiently read the Italian text and enthusiastically suggested its translation. Bob Walters and Tess Kissinger put me in contact with Bob Sloan, who coordinated the editing process at Indiana University Press; Barbara Ferrett Rogers translated the whole text with passion; Simone Maganuco made most of the updates for the English edition; Mami Azuma, Cesare Brizio, and Daniela Cipollone impartially offered their wise advice.

I cannot overlook mentioning all those people who, by their discoveries or their professional expertise in excavating, preparing, restoring, and casting the fossil finds, have made possible the enhancement and study of Italian dinosaurs and other fossil reptiles: Debora Affer, Flavio Bacchia, Vittorino Cazzetta, Luciano Chemini, Piero Conca, Federico Confortini, Claudio del Prato, Piero Fanchin, Fabio Fogliazza, Amedeo Giani, David Govoni, Lorenzo Magnoni, Matteo Malzanni, Ferdinando Moretti Foggia, Giovanni Pasini, Onofrio Pepe, Sergio Rampinelli, Luis Rey, Damiana and Salvatore Santoro, Riccardo Sigismondo, Silvio Tassinari, Giovanni Todesco, and Angelo Zanella. A peculiar thought to the memory of Ilario Sirigu.

My thanks go to the Paleontology Commission of the Ministry for

Cultural Heritage and Activities, in the person of Luigi Capasso, and all the Archeological Superintendences that have kindly granted authorization for publication of the iconographic material, in the persons of Giuliana Tocco, Maria Fariello, and Leonardo Vitola (Salerno); Franca Maselli Scotti, Alvaro Colonna, and Franco Bocchieri (Trieste); Angelo Maria Ardovino, Raffaella Poggiani Keller, and Elena Menotti (Milan); Giuseppe Andreassi (Taranto); and Giovanna Gangemi (Padua). My special thanks to the Director of the Museum of Natural History of Milan, who authorized the use of several pictures from the archives, and to Marco Avanzini of the Tridentine Museum of Natural Sciences for the Independent Province of Trento.

A word of mention goes to Macmillan Magazines, which granted the rights to reproduce the *Nature* essay; the editorial staff of *Oggi* (Paolo Occhipinti, Pino Aprile, and Franco Capone), thanks to whom the dinosaur of Pietraroia was returned to the competent authorities; the companies Fossilia, Salnova, Stoneage, and Tridimont; and the Natural Science Museum of Bergamo.

Last, I should like to thank the Radiology Department of the Ospedale Maggiore of Milan, in the persons of Giorgio Cosmacini, Daniela Castagnone, Andrea Lovaria, Giovanni Tafuni, and Giorgio Sergio, for their help in carrying out the x-ray examinations of the Besano fossils.

The photographs published in this book are by Marco Avanzini, Cristiano Dal Sasso, Giovanni Pasini, Isabella Premoli Silva, Luciano Spezia, Giorgio Teruzzi, Franco Valoti, and Leonardo Vitola. The illustrations are by Fabio Fogliazza, Tullio Perentin, and Luis Rey, and the computer graphics are by Simone Maganuco, Franco Nodo, Claudio Pagliarini, and Marco Yais.

C.D.S.

# INTRODUCTION

THERE HAVE ALWAYS been dinosaurs in Italy, but we simply didn't see them. For the experts who study the most famous reptiles in history, Italy had been an odd exception for almost two hundred years. While in the rest of Europe dinosaurs emerged just about everywhere, in Italy there was not even the slightest trace of them. And that was not because there were no paleontologists.

Centuries ago, at a time when studying fossils was not thought of as a scientific activity, Italian naturalists recorded the fundamental stages in the history of paleontology, to the point of being rightly considered today as the founders of this discipline. An enormous number of Italian fossils were studied and made famous throughout the world by enlightened researchers, but the great majority of the Mesozoic vertebrates were represented by marine animals. As the geologists confirmed, the rocks that were formed during the "era of the reptiles" could never have contained dinosaur bones because they were formed on an ocean bed.

The swift succession of discoveries that belied this idea was as anomalous as the prolonged lack of evidence. If we exclude the first two areas in which tracks were discovered, all the traces of the presence of dinosaurs in Italy have come to light in the past ten years. And they are so many and so interesting as to make it possible to write a book about them. This was the common intuition that three years ago led me to make the acquaintance of the journalist Giuseppe Brillante, to whom I should like to express my special thanks, first for having incited me, and then for having helped me draw up the texts and adapt them into a more comprehensible form. A common intent achieved with a different approach, but precisely for this reason even more fruitful: interest in technical data and their scientific importance, on the one hand, and passion for novelty and its impact on the public, on the other. Without mediation of this kind, paleontological discoveries, as sensational as they may be, are heavy going and difficult to digest for some.

This book, now updated in its English edition, has consequently been created with the serious intention of making it intelligible to all, so as to prevent the memory of an important moment for Italian paleontology from sinking into oblivion. Behind the scenes of the sensational scientific discoveries that are being made in Italy there are dates, names, and human stories that would otherwise remain unknown to the official records, and which we believe are worth being told.

Italy is an opencast museum. It is the country with the richest history but nonetheless with a rich prehistory. Its artistic and naturalistic treasures are great, as are the problems posed by their protection. This book also aims to arouse the attention of all those who administer Italy's cultural heritage, to enable them to fully understand the importance of scientific enhancement and public promotion of her paleontological heritage at both the national and international levels. The discovery of dinosaurs in Italy is a wonderful occasion for scientific research and popularization in this part of the world. Now that the raw material is here—and is of excellent quality—the American *T. rex* and the *Velociraptor* won't seem such an unattainable dream. The whole world admires the little *Scipionyx,* the Similaun man, and numerous other finds due to their exceptional state of preservation. Mother Nature has done an excellent job. Will we be able to preserve and protect them the way she has?

C.D.S.

# DINOSAURS OF ITALY

# 1

# DINOSAURS

THE DAWN OF TIME has rendered tangible evidence of the existence of fantastic, sometimes gigantic, and, for some, monstrous animals which have now disappeared from the face of the earth. We know from their fossils that dinosaurs really did exist, and that they lived on this planet before the appearance of mankind.

The word "fossil" derives from the Latin verb *fodere,* which means to dig. In fact, fossils are the remains and traces of organisms that have lain preserved, mainly petrified, in sediments accumulated on the earth's surface, and which have been exposed by natural erosion of the sediments or by excavation.

The veracity of the fossils is unquestionable. Up until the sixteenth century, fossils were considered to be "freaks of nature" that happened by pure chance to resemble "monsters," or living creatures modeled in the rock by some mysterious "plastic force" or by phenomena such as lightning. (Only Leonardo da Vinci guessed their real meaning.) Later, a number of exceptionally well-preserved fossil discoveries began to provide overwhelming proof of their real nature. Today, techniques such as chemical analysis enable us to unequivocally ascertain their organic origin.

The problem, if any, has been acknowledging that the fossils date back to extremely remote past eras—not to several hundred but to millions of years ago. This is an idea which, as we will see, was not easily accepted by either men of science or the man on the street.

## The Dinosaurs' Timepiece

According to one eighteenth-century divine, Archbishop Ussher, the world was thought to have been created at 9:00 A.M. on October 26,

In the past, fossils were thought to be freaks of nature generated by mysterious forces within the rocks. But the perfection of certain findings, such as this sole (*Amphistium paradoxum*), which lived 48 million years ago in the province of Verona (Bolca), persuaded even the most stubborn skeptics to change their minds. *Photo by C. Dal Sasso.*

# 3
# Dinosaurs

4004 B.C. Ussher inferred this date chiefly from his study of the Bible by means of complicated calculations based on the succession of the generations quoted in the Holy Scriptures, using the best historical and astronomical records of his time. Geological studies and the observation that phenomena such as valleys gouged by rivers and coastlines shaped by the ebb and flow of the sea required long periods of time led to the conjecture that the earth was much older. The nineteenth-century discovery of the mechanisms of biological evolution contributed greatly toward changing the general opinion, since the modification of one species into another, which was documented precisely by the fossils, required much longer periods of time. Nevertheless, even as late as the beginning of the twentieth century, the world was believed to be much younger than we know it to be today. In 1918, the American paleontologist Henry Fairfield Osborn maintained that the Paleozoic Era began 30 million years ago, whereas today we know that it began approximately 540 million years ago.

How can we be so certain of these dates? What makes our knowledge so advanced compared to that of our predecessors? Thanks to the discovery of the structure of the atom, we are now capable of measuring the absolute age of rocks by utilizing the natural radioactive decay of certain minerals.

Before the discovery of the atom, geologists could assign only relative dates to rocks. The only criterion for assessing the age of a rock (and of a fossil embedded in it) was its stratigraphic position: one sediment was older than another one if it was underneath it, because it was deposited first. This is why, when we speak of geological eras, we still use adjectives such as "lower," "middle," and "upper"; these are spatial and not temporal attributes.[1]

At the beginning of the twentieth century, the English physicist Ernest Rutherford discovered that the constituents of all matter, namely atoms, have an infinitely minute mass made up of a nucleus of protons and neutrons and a swarm of electrons that revolve around them like satellites. With the cooperation of his colleague Hans Geiger, Rutherford also discovered that when they are not in perfect dynamic equilibrium, some of these particles can abandon the atom and produce radioactivity. Radioactivity is a spontaneous loss of mass that occurs in certain unstable atoms. This transformation, referred to as decay, takes place at a constant rate. By knowing the decay time of an atom and measuring how much of its mass remains in a rock, it is possible to know when that rock was formed, to measure its age. The only problem is that the "right" radioactive elements (uranium, thorium, potassium, rubidium) do not occur in sedimentary rocks, which are the fossil-bearing ones, but in igneous rocks, which do not usually contain fossils. In order to date a dinosaur bone, it is necessary to find a stratum of volcanic products or an igneous intrusion just a little above or below the stratum and then estimate its difference in age.

Since radiometric dating came into use, geology and paleontology have taken great strides forward. We now know that our planet is no less than 4.5 billion years old, that the dinosaurs were not wiped out by the universal deluge a few thousand years ago but are much older, and that the first hominids are no less than 6 million years old.

The dinosaurs represent one group of terrestrial animals that has enjoyed the greatest success in the history of life on earth. Dinosaurs dominated all the continents for 165 million years, starting from approximately 230 million years ago until they were all but extinguished 65 million years ago. For a human being, who rarely lives longer than 100 years and who spends his days in a frenzied and narrow period of time that he measures in hours and minutes, it is difficult to grasp what 165 million years means. When compared with the length of the era of the dinosaurs, the entire existence of mankind is equivalent to the twinkling of an eye.

To give an idea of the duration of the geological eras, let us take a moment to imagine that the 4.5 billion years of the history of the earth have been compressed into a 24-hour day. According to this hypothetical geological clock, an hour would be equivalent to 187 million years, a minute would be 3 million years, and a second would be 50,000 years. Within the space of the day, we would see our planet born at 0 hour and remain devoid of any form of life until midday. The first multicellular animals would not appear until just before 9:00 P.M., whereas the dinosaurs would hold sway from 10:46 P.M. to 11:40 P.M. It is somewhat perturbing to discover that, in this bewildering dimension of time, the first human would not appear on the earth until 38.4 seconds before the stroke of midnight.

In real time, the dinosaurs disappeared practically 60 million years before the hominids appeared. Cavemen would never have been able to come face to face with one. And considering how many million years separate us from the last dinosaurs, despite a number of modern legends, none of them can have survived to our day and age. If we really want to look for living dinosaurs, we need only look to the skies: the only dinosaurs that we can see in the flesh are their descendants, those multicolored feathered animals that we are accustomed to calling birds.[2]

## What Is (and What Is Not) a Dinosaur

The term "dinosaur" derives from the Greek words *deinòs* (fearfully great) and *sàuros* (lizard, reptile) and was coined in 1841 by the English naturalist Sir Richard Owen to identify the great reptiles of the past whose gigantic bones had begun to emerge from excavation sites almost everywhere in Europe. The term very quickly came into common use (and abuse), to the extent that today it is still very often imprecisely used as a synonym of "large extinct animal" to indicate a variety of fossil vertebrates that have nothing to do with dinosaurs. The "false dinosaurs" most widely mentioned are the flying reptiles (pterosaurs) and a medley of marine reptiles that are wholly unrelated to one another (ichthyosaurs, plesiosaurs, mosasaurs), all contemporaries of the dinosaurs but adapted to totally different lifestyles. Even the mammoths and other large mammals that lived in much more recent geological periods are very often mistaken for dinosaurs.

So what, then, is a dinosaur?

In classifying living creatures, scientists came to realize that in nature there are anatomical features that are widespread among certain organisms. These features can be found in very ancient animals and are used to identify the emergence of new types of animals. Vertebrae, for example, are shared by numerous animals (which, not by chance, are referred to as vertebrates) and help define a basic lineage, one that appears early in the evolutionary time scale.[3] A dinosaur is undoubtedly a vertebrate. It is a multicellular animal provided with an articulated internal skeleton, a supporting framework the axis of which is composed of a chain of bones referred to, precisely, as vertebrae.[4] As we approach the evolutionary maze that leads to dinosaurs, we may enter the opening marked "vertebrates."

A short way down the path, we must choose between two fundamental structural levels: one that exhibits four limbs and one that does not (fishes). Although some dinosaurs were bipedal, there is no doubt that they all had four limbs. Therefore, we choose the path marked "tetrapods." One step further, we leave the amphibians to one side and opt for the great group of amniotes, which are tetrapods that lay eggs

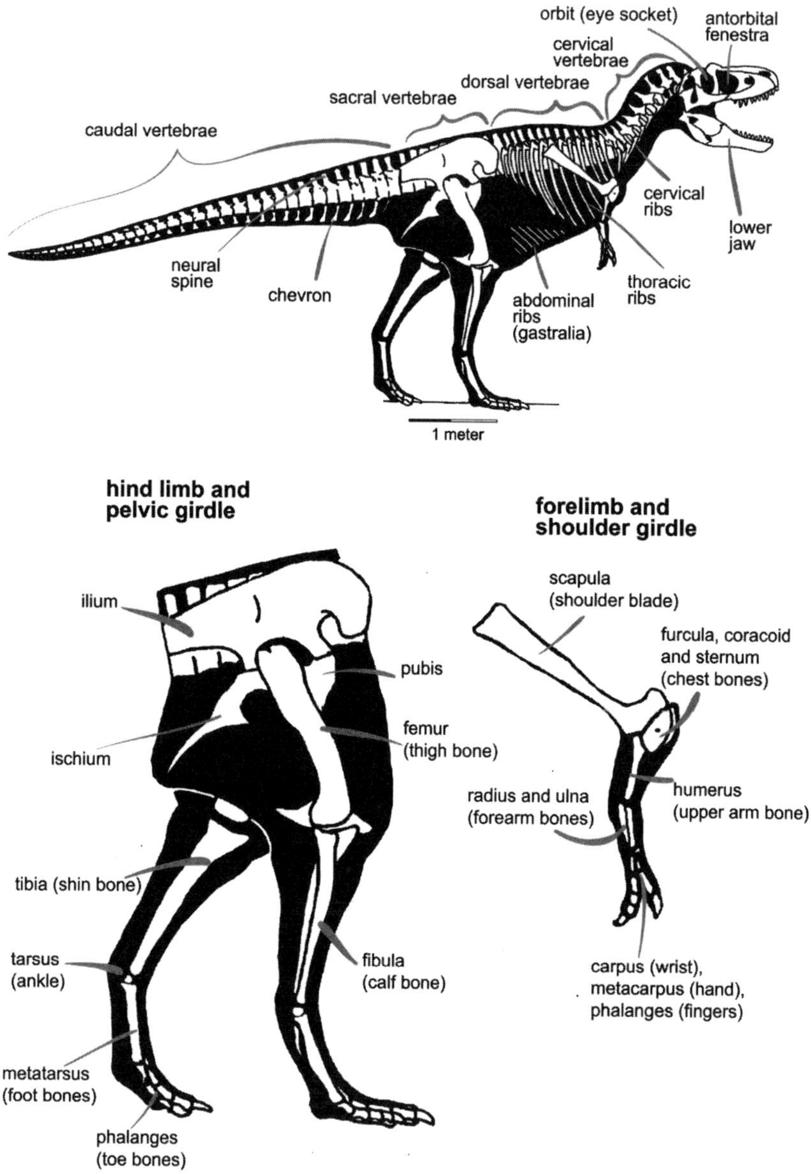


Like all vertebrates, dinosaurs (in this case a tyrannosaur) have an internal skeleton whose supporting axis is the spinal column. In comparison to the approximately two hundred bones making up the human skeleton, dinosaur skeletons have around three hundred. The number varies primarily in the tail. *From Farlow and Brett-Surman 1997, modified. Graphics by F. Nodo.*

provided with an amniotic sac and covered with a hard shell. The numerous discoveries of fossil dinosaur eggs confirm that the choice is right: dinosaurs are amniotes.

Unfortunately, many of the characteristics that help us distinguish between living amniotes (reptiles, birds, and mammals), such as type of metabolism, method of reproduction, and presence of scales, feathers, or fur, are not useful in recognizing their fossils, since these are attributes that are almost never preserved. Instead, paleontologists have had to painstakingly study the differences at the skeletal level. Therefore, the matter is more complicated. Now we have to choose between the amniotes that have an opening beneath the temporal region of the skull (synapsid reptiles) and those with a skull that has two temporal openings (diapsid reptiles) or has none at all (anapsid reptiles); since dinosaurs have two temporal fenestrae, there is no doubt that they are diapsid reptiles.

We have nearly entered the hall of the dinosaurs. In front of us we have the lepidosaurs (lizards and their kin) and the archosaurs. The archosaurs are distinguished by the presence of an antorbital fenestra (an aperture situated between the bones of the skull, in front of the orbits for the eyes). This fenestra is more than evident in the skulls of all dinosaurs, and so we can conclude that dinosaurs are diapsid archosaurian reptiles.

But when all is said and done, what makes dinosaurs such extraordinary animals and so very different from all the other reptiles? It can't possibly be that the apertures in their skulls, which for that matter crocodiles and birds also have, are the reason for the dinosaurs' success. There must be something else. First and foremost, unlike flying and aquatic reptiles, dinosaurs were decidedly terrestrial reptiles. They certainly did not live that improbable amphibian life immortalized in artistic representations and exhibited in museums up to the early 1970s. Dinosaurs, at least understood in the strict sense of the word, were not even flying reptiles.[5] They were made for walking over the ground and were so well adapted for this that they represent a model of extraordinary architectural perfection.

To distinguish a dinosaur from any other reptile, you need only look at it from the front and observe its limbs. The most significant difference, in fact, concerns the position of the limbs with respect to the body. In a dinosaur, the hind limbs are directed vertically beneath the body. The joints are aligned (the so-called erect posture, which enables the animal to maintain its body raised well above the ground) and do not protrude at the sides, as is the case in the great majority of reptiles (the so-called "sprawling" posture typical of lizards).

Dinosaurs inherited this posture from a group of archosaurs, informally known as "thecodonts" but not closely related to one another, and in particular from a type of animal identified today by the name of euparkerid. Like the dinosaurs, the euparkerids possessed an antorbital fenestra, and above all they showed a tendency to gradually shift the position of their limbs from the side to a position beneath their bodies and at the same time to reduce the angulation of their leg joints. Around 245 million years ago, animals similar to *Euparkeria* branched off into two evolutionary lines: one was to lead to the crocodiles, and the other to the pterosaurs (flying reptiles) and the real dinosaurs. A feature that these two latter groups had in common was a joint between limb and foot, called the advanced mesotarsal joint, which allowed greater flexing of the ankle bones.

As far as we know today, the most probable ancestor of the dinosaurs is *Lagosuchus,* a very slender-built animal about 50 centimeters long with rear limbs longer than the front ones, which lived during the Middle Triassic in Argentina. In fact, a similar anatomical structure can be found in the oldest known dinosaur, which dates back to approximately 230 million years ago, and which lived in the same geographical area. *Staurikosaurus,* discovered in the formation of Santa Maria del Rio Grande do Sul in Brazil, must have been a small and agile bipedal carnivorous dinosaur.[6]

Despite the "prehistoric" appearance of some present-day reptiles, most are not direct descendants of the dinosaurs. This can readily be seen from how they move: they sway their body left and right, almost sliding on their bellies.[7] From an evolutionary standpoint, this system of locomotion is very primitive (it was inherited from the swimming movement of fish), and it may be the reason why lizards and crocodiles have never dominated our planet as dinosaurs did.

Paleontologists use many anatomical features related to the columnar (erect) posture of the limbs to recognize dinosaur bones from among the fragments that come to light during their excavations: a sacral bone composed of three or more fused vertebrae, a pelvis showing an open cavity (acetabulum) to allow the insertion of the thighbone, and a thighbone with a head oriented toward the inside of the hip are all typical features of the skeletons of dinosaurs, indicating that the limbs were positioned vertically beneath the body.

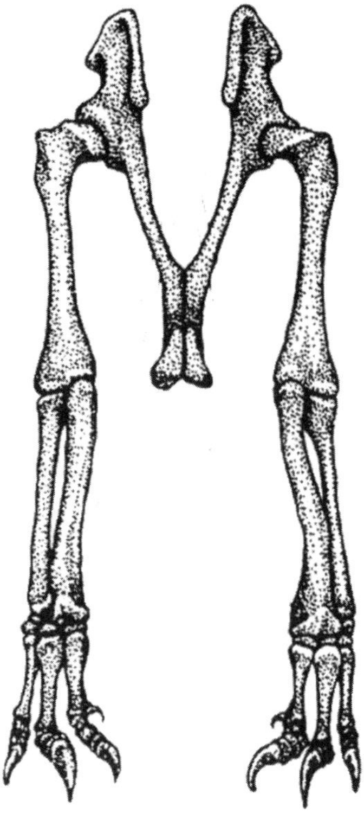

DINOSAURS

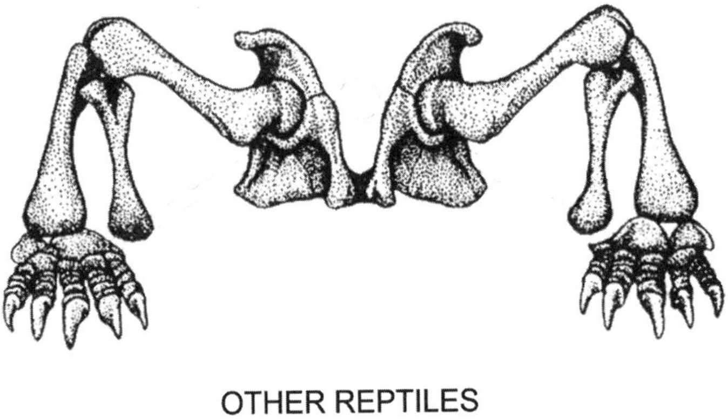

OTHER REPTILES

Dinosaurs differ from other reptiles in the columnar posture of the limbs. Note, in particular, the different orientation of the femur (vertical in the dinosaurs and almost horizontal in the other reptiles). *From Dal Sasso 1995, modified. Graphics by M. Yais.*

# 10
## Dinosaurs of Italy

The columnar posture and relative rigidity of the spinal column are among the reasons for the dinosaurs' success on dry land, because this architecture is capable of supporting enormous weight and at the same time offers a certain amount of agility, and in some cases it permits bipedal locomotion, whereby it is possible to use the forelimbs for other purposes. That posture is also one of the main reasons for their failure in water, because it prevents "anguilliform" movements—the sideways undulation necessary for swimming. This does not mean that, when the need arose, dinosaurs could not go into the water and even swim, but in any case that was not their natural habitat.

Now we have a clearer idea of the distinction between dinosaurs and marine reptiles. All the reptiles that inhabited the seas during the era of the dinosaurs originated from terrestrial animals similar to lizards. Those most well adapted to living in water, the ichthyosaurs, also retained the lateral undulating movement used by their ancestors to travel over dry land.

Very often evolution almost retraces its steps, and hence after the widespread extinction at the end of the Cretaceous Period, the seas left behind by the reptiles were reconquered by other great terrestrial animals: some mammals, distant cousins of the deer and the cow, changed their legs into flippers and gave origin to cetaceans (dolphins and whales). Many people, deceived by the similarities in body structure, believe it is logical that dolphins descend from ichthyosaurs, whereas in actual fact this is a phenomenon that biologists call adaptive convergence: different, totally unrelated animals assume similar shapes to be able to live in the same environment. To live in water, you have to know how to swim, and so, again and again, fishes, reptiles, and mammals, at different times in the history of life, "invented" fins.

The same thing happened in the air: insects grew wings more than 120 million years before the reptiles, during the Carboniferous Period. During the era of the dinosaurs, the pterosaurs (the so-called flying reptiles) developed membranes of skin stretched between the forelimbs and hind limbs, which reached a wingspan of no less than 13 meters. When they disappeared, the skies were already inhabited by birds, which in the meantime had developed from certain small carnivorous dinosaurs and not from the pterosaurs. And after the birds, mammals also attempted to conquer the skies, with bats.

Another myth to be disproved is the idea that dinosaurs were all enormous. Many carnivores were shorter than a human, and some were

*Lagosuchus* (*top*) is a "thecodont" archosaur considered to be the most probable ancestor of the first dinosaurs. *Lagosuchus* and *Eoraptor* (*bottom*) were both bipedal and measured no more than a meter in length. *Model by J. Blanco. Photo by C. Dal Sasso.*

even smaller than a chicken. One of the reasons why in 1981 Giovanni Todesco, the discoverer of *Scipionyx,* did not realize that he had found the first Italian dinosaur was precisely because it was a little animal only 40 centimeters long. "Ciro" died shortly after birth, but even if he had grown to be an adult, he would not have been more than 1.5 meters long.

## Evolution of an Idea

One day in the spring of 1822, or so the story goes, in the hills of the county of Sussex in England, Mary Ann Mantell noticed two strange objects in a pile of stones dumped at the side of a road. Just two years later, the jaw of a large carnivorous reptile, called *Megalosaurus,* was described and acknowledged as the first dinosaur fossil by another Englishman, the Reverend William Buckland. The fossils discovered by Mrs. Mantell were so unusual that her husband Gideon, a doctor with a passion for the earth sciences, after having studied them at length, presented them in 1825 to the Royal Society of London as the teeth of a "giant iguana." *Iguanodon* thus emerged from the past with a misleading name. The iguanodon and the carnivorous dinosaurs, such as the megalosaurs, were depicted in prints of the period as gigantic lizards that dragged themselves awkwardly along on four feet, crawling rather than walking.

They were reconstructed, life-size, as massive quadrupeds but with their bellies raised from the ground, by Benjamin Waterhouse Hawkins in 1854 and placed on the grounds of the Crystal Palace in London. Thanks to the pioneer studies of Richard Owen, the inventor of the term "dinosaur," the "fearfully great lizards" were recognized as having a vertical posture of the limbs. Rather than a lizard with a strange horn on its nose, the iguanodon now appeared to be an enormous rhinoceros covered with scales and armed with deadly talons.

In 1878, a whole herd of iguanodons—no fewer than thirty-three skeletons, many of them exceptionally complete and articulated—were unearthed in a coal mine at Bernissart, in Belgium. It was then that Louis Dollo, a young paleontologist from the Brussels Museum, realized that the horn was in actual fact the tip of the thumb, and above all he realized that these animals could stand upright, supporting themselves on their hind limbs. The living model used, however, was inaccurate. The dinosaurs of the end of the nineteenth century and the first decades of the

The progress made by paleontologists on the subject of dinosaurs has given rise, in less than two hundred years, to radical changes in the reconstruction of their appearance. Four different interpretations of *Iguanodon* are illustrated here: (A) according to Mantell and Nibbs (1835–1838); (B) Owen and Hawkins (1852–1854); (C) Dollo and Parker (1880–1950); (D) Lambert (1989 to date). *From Rosenberg and Wolberg 1994, modified. Graphics by M. Yais.*

twentieth century stood upright like kangaroos, with their backs far too erect.

In 1940, Rudolph Zallinger painted an enormous mural for the Peabody Museum at Yale portraying the dinosaurs as slow-moving, massive reptiles, standing erect but awkwardly resting on fat tails. The huge painting became so famous that it influenced all the iconography for much of the century. And yet the dinosaur tracks that came to light did not reveal the deep groove that would have been dug by a tail dragged along the ground in that way. The iguanodon did not raise its tail until thirty years later, following ichnologic research and bioengineering and comparative anatomy studies carried out on birds, the living descendants of dinosaurs.

If we wish to give a more precise date to the birth of the modern conception of dinosaurs, we have to wait until 1969. That year, the American paleontologist John Ostrom, of Yale University, published the description of *Deinonychus* (just for the sake of clarity, the American cousin of the Asian *Velociraptor*): its anatomy unmistakably revealed that it was an agile, fast-moving animal, more similar to a bird than to a modern reptile. *Deinonychus*'s weapon of attack was an enormous scythe-like talon situated not on the forelimbs but on the inner toes of its feet. To rip open its prey, this dinosaur thus had to jump or support itself on one foot. A tail reinforced by a bundle of bony rods strengthened the hypothesis of an acrobatic dinosaur. The discovery of *Deinonychus* paved the way for Robert Bakker, at that time considered the *enfant terrible* of paleontology. In 1975, Bakker, who acquired his experience under the guidance of Ostrom, came to the conclusion that it was a mistake to compare dinosaurs with modern reptiles. Numerous paleontological data indicated that dinosaurs were capable of continuous activity and complex behaviors similar to those observable today in warm-blooded animals. This gave rise to a debate about the physiology of dinosaurs that, as we will see later, is anything but concluded.[8]

Recently, a more correct interpretation of the "horseshoe-shaped" tracks and the structure of the pelvic girdle of certain vegetarian dinosaurs brought the iguanodon back with its feet (or rather, its hands) on the ground. The robust sole of the front extremities gathered into a pad in the shape of an upturned U, the thick tail served as a balancer, and the saddle-shaped pelvis made *Iguanodon* a "facultative biped"; that is, all of these features enabled it to adopt a bipedal or quadrupedal posture, according to the circumstances.

At the end of the 1970s, the discovery of *Deinonychus,* a predator capable of leaping agilely on its hind limbs, brought an end to the idea that dinosaurs were slow, clumsy reptiles. *Photo by C. Dal Sasso.*

Although paleontology is a relatively young science, ideas about dinosaurs have changed many times in 180 years. Now, at the beginning of the third millennium, it would seem that our idea of dinosaurs is about to undergo a further and even more radical revolution: many modern dinosaur reconstructions wear a coat of feathers. But we'll talk about that later.

## Who Gave Dinosaurs Their Names?

How do paleontologists create the names of the dinosaurs they discover? Where do such complicated names spring from, and why are they in Greek and Latin? The matter is anything but banal. The names are invented by the discoverers or the experts who are the first to realize that they are looking at a new species. However, there are precise rules to be followed.

Not only every dinosaur, but every animal or plant species has to be identified by a binomial (a sort of name and surname) according to the standards of the International Code of Zoological or Botanical Nomenclature.[9] There is even a committee of experts, based in London, that examines controversial cases, as when it is called upon to decide which name should be recognized as valid when two or more names have unwittingly been assigned to the same animal. In one such question of "priority," the committee determined that the animal known as *Brontosaurus* should be renamed *Apatosaurus.* The committee had discovered that the paleontologist Marsh used the latter name in 1877 and the former in 1879 to distinguish the remains of two dinosaurs that later, thanks to the discovery of more complete skeletons, were recognized as belonging to the same genus.

Only the binomial unmistakably identifies a species; it must be written in Latin (there have been exceptions in recent years) and in italics, with the first name starting with a capital letter—which may also be abbreviated—and the second name always in lowercase letters and never abbreviated. For example, the most famous species of tyrannosaur is designated *Tyrannosaurus rex* and not Tyrannosaurus Rex; *T. rex* and not T-rex. The triceratops is *Triceratops horridus* or *T. horridus.* The first name identifies the genus, a group composed of one or more closely related species (there are *Triceratops horridus* and *Triceratops prorsus;* there are *Homo sapiens* and *Homo erectus*); the first name combined with the second—but not only the second—identifies the species. *Horridus* (with a capital initial letter) or *horridus* without the name of the species does not mean anything because it is an adjective and as such can be combined with other animal or plant genera (*Moloch horridus,* for example, is a spiny lizard that lives in the Australian deserts).

But where do these strange and sometimes hard-to-pronounce names come from? When a scientific name is created, it is not chosen haphazardly. The idea is to choose one or several words, usually descriptive but sometimes honorific, translated into classical Greek or Latin, which can somehow commemorate the discovery. Very often two meanings are combined into a compound word. Consequently, there are various possibilities to choose from, according to whether it is preferred to highlight the site of the discovery (e.g., *Besanosaurus* to specify that it is a "reptile from Besano"), an anatomical characteristic (e.g., *Anatotitan* to indicate the presence of a "duck's bill"), a lifestyle (e.g., *Velociraptor* must have been a "swift predator"), or a personality to whom to dedicate the new discovery (e.g., *Lambeosaurus* in honor of L. Lambe). In recent years, there have also been "sponsored" dinosaurs, new species whose scientific names recall the trade names of the company that financed the excavations or the research that enabled their discovery (e.g., *Atlascopcosaurus, Qantassaurus*).

Since it is difficult to associate fossil tracks with the exact species that produced them, the tracks are usually given a different name. In the past, attempts were made to devise a name that recalled the genus of dinosaur that was thought to have left them, but the numerous erroneous interpretations created even greater confusion. The footprints of a South American predator were called *Hadrosaurichnus* because they were initially attributed to a duck-billed dinosaur, and those of a large Australian carnivore, in no way related to the tyrannosaur, were given the name *Tyrannosauropus*. The endings-*ichnus* and -*pus* in ancient Greek mean "track" and "foot" and are therefore common among the names of fossil footprints, just as the ending -*saurus* (reptile, lizard) is common among the names of fossil and present-day reptiles.

The passing of time is also marked by conventional names. The history of the earth has been divided by geologists into numerous intervals of time. These divisions, referred to as geochronological units, were made by studying the stratified sequences of sedimentary rocks and the fossil content when proceeding from one sequence to another. The classic example is the separation between the Mesozoic Era and the Cenozoic Era, popularly known as "The Age of Dinosaurs" and "The Age of Mam-

*Allosaurus* has changed names three times due to synonyms (*Epanterias, Antrodemus*) attributed to specimens that seemed to be different and were later recognized as belonging to the same genus. © *Natural History Museum of Milan. Photo by L. Spezia.*

mals," the boundary being marked by the disappearance of the great reptiles.

Each era is subdivided into periods, and these in turn are divided into epochs, stages, and substages. The names of the subdivisions of the history of the earth have the most diverse origins. Some derive from the localities where the lands that formed during that period were identified for the first time (Jurassic takes its name from the French region of the Jura); others derive from geological features (the Carboniferous Period is so called because of the abundant layers of fossil carbon present in the rocks of that period).

Some subdivisions of the history of the earth have Italian names. These are intervals of the Tertiary and Quaternary Periods, which take their names from regions in which the sediments of these ages are widespread; hence we have the Astian, Piacenzian, Messinian, Tortonian, Emilian, and Calabrian stages.

Intelligence induces human beings to create order in things and consequently to classify everything: plants, animals, times, and places. In this way we can gain a better knowledge of the world around us. And so, for example, it comes naturally to us to give a name to every living creature and unite those that are similar into the same group. It is true that, as we said before, in nature there exists a hierarchy in the distribution of anatomical features, which in fact helps us to group species into genera, genera into families, families into orders, orders into classes, and so forth. However, we should not forget that the names we choose to identify these groups (including the word "dinosaur") are pure creations of our mind.

## Starring and Guest Stars

Dinosaurs make up an incredibly diversified group of vertebrates. And yet we know but a small number of the species that inhabited the Mesozoic world, since the individuals that have been handed down to us in the form of fossils are the result of preservation processes that are very rare in nature. We can only imagine how many variations on the theme evolution produced within the various families of dinosaurs, of which very few representatives are as yet known. These gaps, compared with the variety of fossils discovered, have created a great deal of controversy in the classification of dinosaurs. Current opinion is that dinosaurs are a monophyletic group, which means that they have a common ancestor. Nevertheless, paleontologists are still accustomed to distinguishing two major orders of dinosaurs according to the different arrangement of the bones of the pelvic girdle, which apparently reflects an evolutionary divergence that occurred at the beginning of their history. Saurischian dinosaurs, the "lizard-hipped" dinosaurs, are characterized, like other reptiles, by a pubic bone facing forward toward the belly; ornithischian dinosaurs, whose name means "bird-hipped," have a pubic bone facing backward toward the tail, parallel to the ischium.

Saurischian dinosaurs are currently distributed into two suborders; one consists of exclusively bipedal, and for the most part carnivorous, animals (theropods), and the other includes large herbivorous quadrupedal dinosaurs (sauropodomorphs).

Until a few years ago, theropods were divided into two groups (coelurosaurs and carnosaurs), almost exclusively according to their body proportions. The first group included the little svelte-bodied theropods with small skulls and relatively long, slender necks and limbs, while the "carnosaurs" were the large carnivorous dinosaurs with huge heads and front limbs decidedly shorter than the hind limbs.[10] Theropods have turned out to be a much more heterogeneous group with highly diversified forms and adaptations (including feeding habits). For example, the coelophysoids (primitive theropods with four-fingered "hands") include medium to small slender forms with pointed skulls. The ceratosaurs include animals of all shapes and sizes, while spinosaurs have a long snout, more similar to that of crocodiles.

With the important exception of the huge tyrannosaurids,[11] the coelurosaurs were extremely agile dinosaurs that enjoyed great evolutionary success with families such as the ornithomimids, oviraptorids, therizinosaurids, dromaeosaurids, and troodontids. The coelurosaurs lived alongside the other dinosaurs until their extinction at the end of the Cretaceous Period. However, some of them, which in the meantime had developed an insulating covering of feathers and quills that later became useful for flying, survived the catastrophe and continued to evolve into the birds of today. The first dinosaur skeleton discovered in Italy, the famous *Scipionyx samniticus,* to which Chapter 3 is devoted, belongs to the coelurosaurs. A much larger type of theropod was recently discovered in the province of Varese (Lombardy) and is described for the first time ever in this book, in Chapter 5.

Most herbivorous saurischian dinosaurs make up the suborder of the

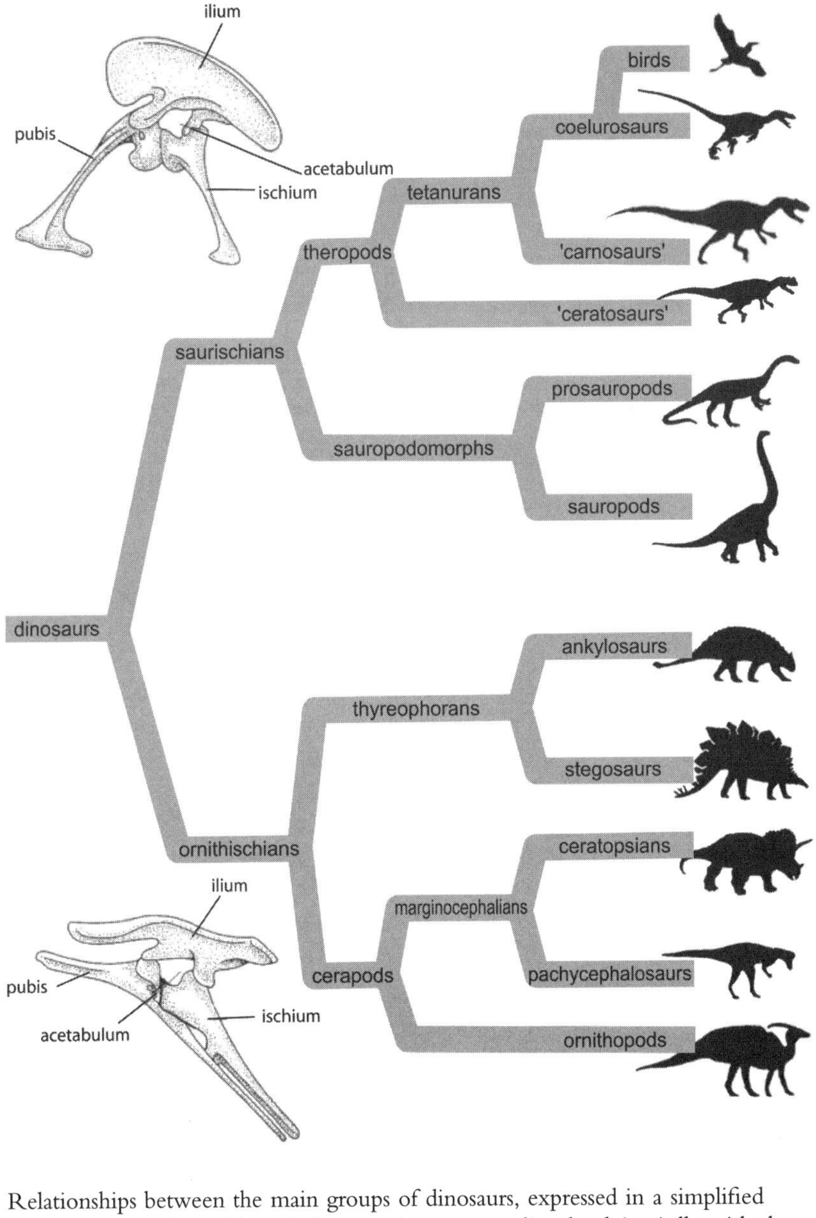


Relationships between the main groups of dinosaurs, expressed in a simplified cladogram. The lizard-hipped dinosaurs have a tri-radiated pelvic girdle with the pubis facing forward. The bird-hipped dinosaurs have a tetra-radiated pelvis with the pubis ending in an anterior ramus and a posterior ramus. *From Lucas 1994, modified. Graphics by F. Nodo and M. Yais.*

sauropodomorphs. They were almost all quadrupedal, with the exception of some members of the prosauropods, their primitive representatives. The infraorder of the sauropods includes the most impressive dinosaurs, the largest and heaviest animals that have ever walked the earth. Sauropods had very long necks and tails. Their limbs were relatively massive and their heads very small. Some of the oldest sauropods left their fossil prints in the area around Rovereto (Veneto) and are among the protagonists of the next chapter.

Ornithischian dinosaurs had teeth specialized for an almost exclusively vegetarian diet, but in terms of variety of forms they were a highly heterogeneous group. The suborder of ornithopods included two quite distinct types of dinosaurs: the most primitive ones, such as the heterodontosaurs and the hypsilophodonts, were small bipedal dinosaurs similar to the coelurosaurs; the more highly evolved ones, which are believed to have originated from the hypsilophodonts, made up the two large families of the iguanodonts and the hadrosaurs. These reached enormous dimensions and could walk on either two or four legs. Chapter 4 is devoted to the hadrosaurs, several skeletons of which were discovered on the outskirts of Trieste (Friuli).

Very curious-looking ornithischians were the pachycephalosaurs, which had developed a robust cranial dome of solid bone perhaps used in head-to-head combat. The "horned dinosaurs," better known as ceratopsians, were equally bizarre: they had a snout similar to a parrot's beak, one or more horns, and a skull whose rear portion widened out into an enormous bony crest. But evolution produced truly fantastic forms when it came to the suborder of the thyreophorans, better known as the "armored dinosaurs": the stegosaurs were provided with vertical plates or long spikes on their back; the ankylosaurs were covered with a shield of plates that converged on the tail to form an enormous club. The tracks of these incredible dinosaurs appear to be present in the vast tracksite of Altamura in Apulia, which was discovered in 1999 and is described in Chapter 2.

The increasingly widespread opinion is that, due to their diversity of forms and geological eras, dinosaurs were a heterogeneous group from the metabolic standpoint as well. The question is still much discussed, but generally speaking, today it is thought that the small theropods and the small ornithopods were warm-blooded animals (or rather endotherms), while the large theropods were likely to have behaved more like cold-blooded animals in that they could sustain a high level of activity for only short periods. The sauropods and the large ornithopods must have had a somewhat accelerated metabolism during the early stage of their life (as the rapid growth of the bones would appear to indicate) and then became "gigantotherms" as adults. In simpler terms, once fully developed, they managed to keep their body temperature constant by the very fact of possessing a large body mass.[12]

As can be inferred from this brief presentation, the diversity of dinosaurs in Italy is well documented but almost certainly still underestimated. In fact, it is highly probable that in the not too distant future, proof of the existence of an even greater diversity will be unearthed, because the history of the "fearfully great" Italian reptiles began a mere five dozen years ago, in 1941, with a footprint. And it is precisely from footprints that our journey toward the discovery of the Italian dinosaurs will begin.

# 2

# ON THE TRACKS OF THE DINOSAURS

ONE COOL AFTERNOON in the spring of 1988, a man was clambering up among the rocks of an old landslide that had broken away from Mount Zugna in medieval times. The surrounding scenery resembled a moonscape. Crumbling boulders—the remains of entire slopes that had slid down toward the valley—had exposed vast layers of Calcari Grigi di Noriglio, a geologic formation of the Lower Jurassic Period. These are the so-called Lavini di Marco, found in the vicinity of Rovereto in the province of Trento, which are mentioned in the Twelfth Canto of *Inferno* in Dante Alighieri's *Divine Comedy.*

Luciano Chemini was a surveyor with a passion for nature and geology, and that day he was in the area on one of his usual excursions. Suddenly his attention was drawn to something strange on the surface of the rock. There before him, in the *colatoio* (a strip of rock surface worn smooth by flowing water) he was climbing, were a dozen or so holes. Some of them were full of earth, and some of them had grass growing in them. At first Chemini thought that the depressions must have been made by shell explosions during the First World War. But this theory fell by the wayside when he noticed another fact. What really made him curious was that the holes were regularly spaced, and many of them were aligned in two rows. That couldn't have happened by pure chance. And then a bizarre idea flashed through his mind: perhaps they were the footprints of some gigantic animal of the past.

## When the Bahamas Were in Trentino

That night Chemini hardly slept a wink, tormented by an incessant thought: "And what if they were dinosaurs?" He scribbled the suspicion that had kept him awake all night on the back of the photograph he took the next day at the site of the enigmatic discovery and then sent it to the Tridentine Natural Science Museum in Trento. A shot in the dark, it was to be confirmed by Michele Lanzinger, then the curator of the museum, and the two paleontologists consulted for their opinion on the matter: Giuseppe Leonardi, one of the world's greatest experts on ichnology (the science that studies fossil footprints), and Giuseppe Muscio, of the Friuli Museum of Natural History. The news, which was later disclosed in a press conference organized at the Museum of Trento, aroused a great deal of comment among the public.

At least twenty tracks were identified in the *colatoio,* which today bears the name of its discoverer. However, it soon became apparent that they represented only a part of what would later prove to be an immense mine of footprints: a book capable of telling the story of life in prehistoric times, the pages of which have not yet all been turned.

The explorations, which were initiated in 1991 by Leonardi and Lanzinger and continued by a larger study team including the geologists Paolo Mietto and Marco Avanzini, are still in progress and each year lead to the discovery of new traces. During the long-drawn-out process of cleaning up, carried out over hundreds of square meters, the paleontologists had to rid the footprints not only of soil and grass but also of the original rock that still filled many of them like a cast. The most important tracks were then photographed from above and "mapped." By laying transparent plastic sheets on the surface of the rock, it was possible to trace the position and shape of each footprint using colored felt pens. This practical mapping method, together with the plaster or silicone casts made of the inside of the best-preserved footprints, made it possible to study the site at a distance and, above all, prevented the loss of data in the event that the footprints were damaged or covered with rubble again. For the first time in Italy, the group working at the Lavini was an all-Italian team. It included young geology students and enthusiasts; thanks to this experience, a number of them have gone on to discover dinosaur tracks in other locations.

To date, several hundred tracks have been discovered at the Lavini di Marco, scattered just about everywhere in the area. They are not all as spectacular, or as clearly defined, as in the "Chemini Colatoio," but there are enough to infer that the dinosaurs were regular visitors to that spot. In fact, in the outcropping Calcari Grigi, the footprints are found on six different levels, over a vertical thickness of 5 to 6 meters, in the upper part of what geologists refer to as "Lower Member," which was

Lavini di Marco (Rovereto). The first "holes" that in spring 1988 aroused the curiosity of Luciano Chemini as they appear today, after the stratum surface was cleaned up. The track was later called ROLM 9 and is attributed to a large ornithopod. © *Tridentine Natural Science Museum. Photo by M. Avanzini.*

deposited between the Hettangian and the Sinemurian (the first two stages of the Lower Jurassic Period), or between 200 and 195 million years ago.

To understand what happened at Rovereto during that period, we must take a step back in time and try to imagine what this area must have been like in the Lower Jurassic. Where the mountains now stand, there extended a vast tidal flat. This was a white or yellowish muddy strand flat, formed by the deposition of layers of sand and calcareous ooze produced by the disintegration of shells and other skeletal parts of various marine organisms or by chemical precipitation of carbonates in the warm tropical waters. The environment was comparable to shallow marine situations of the present-day Bahamas or certain areas of the Persian Gulf. The coastal waters blended from pale blue to green, depending upon their depth. Here and there you could catch a glimpse of patches above sea level, strips of land and islands partially overrun by vegetation, and inland lagoons and small streams that flowed down to the sea.

The environment was ruled by the slow, steady ebb and flow of the tides, which from time to time submerged or exposed vast mudflats capable of connecting the land masses for prolonged periods. The dinosaur tracks are preserved in intertidal to supratidal levels—that is, in sediment that accumulated just above the mean-tide level. When the water receded, it encouraged the formation of carpets of algae on the muddy surface, and with the sediment still covered by a fine layer of water, the dinosaurs arrived. However, it took very little for everything to change again and for the boundless muddy expanses to return below sea level. Storms, hurricanes, and a further seaward shifting of the shoreline drove the dinosaurs elsewhere, and the cycle began all over again.

The rocks are tangible proof that these tidal flats sloped down westward to the area that is now Lake Garda, in a strait that geologists refer to as the Lombard Basin. Toward the east, where the upland plains of Folgaria, Lavarone, and Asiago are now located, there was probably dry land, because in those localities the Lower Jurassic deposits contain fossil pollen, the only remaining witnesses of the lush vegetation that must have colonized the then-flat Trentino region. The climate was also very different from what it is today: there was no summer or winter. It was a never-ending warm spring, with a temperature constantly ranging from 25 to 27 degrees Celsius.

It was in this environment that the dinosaurs roamed. In all probability they took advantage of the long, muddy banks exposed by the tides to cover the distances between one island and another, perhaps even crossing the deeper reaches by swimming. In fact, even though they were terrestrial animals, and in spite of the gigantic dimensions of some, it is almost certain that they knew how to swim.[1]

At any rate, as they traveled along the tidal flats, their feet sank into the soft muddy carbonate ooze, leaving impressions of their footprints. What happened later is nothing short of amazing. How is it possible for such an ephemeral and temporary thing as a footprint to have been preserved for such a long time?

Only a small percentage of the tracks left by an animal found the conditions suitable for their fossilization. For a footprint to be preserved, it must remain exposed to the air long enough for it to harden and for the sediment inside it to undergo those chemical processes capable of binding one grain of mud to another.[2] Under certain conditions, a few weeks are enough for this to take place, rendering the footprint more resistant to the erosive action of the tides. In the Bahamas today, just as in Rovereto during the Jurassic Period, exposure to hot sunshine causes rapid evaporation of the water contained in the calcareous sludge and facilitates its cementation. Later on, when the level of the water rose, more sediment was deposited on the hardened footprints, creating an overlying layer capable of preserving them forever. However, in many cases, the impression that we see on the rock today is not the one left directly by the animal in the sludge but an "underprint" caused by the pressure of the animal's foot on the layers beneath the one in which it was impressed by direct contact.

## The Sherlock Holmes of Prehistoric Times

Like real detectives, experts interrogate the impressions and obtain many clues from them. As Richard Swann Lull of Yale University once said, they are fossils of living creatures, while skeletons are fossils of dead organisms. We could add that there is a greater probability of finding fossilized footprints than bones. It is obvious that during its lifetime a dinosaur possessed only one skeleton, while it undoubtedly impressed thousands of footprints. In addition, the tracks provide information on the behavior of their makers that cannot be obtained from skeletal remains. Very often, footprints are the only proof of the presence of dinosaurs in certain areas and provide important information on the associations of animals that lived in those areas and on the characteristics of

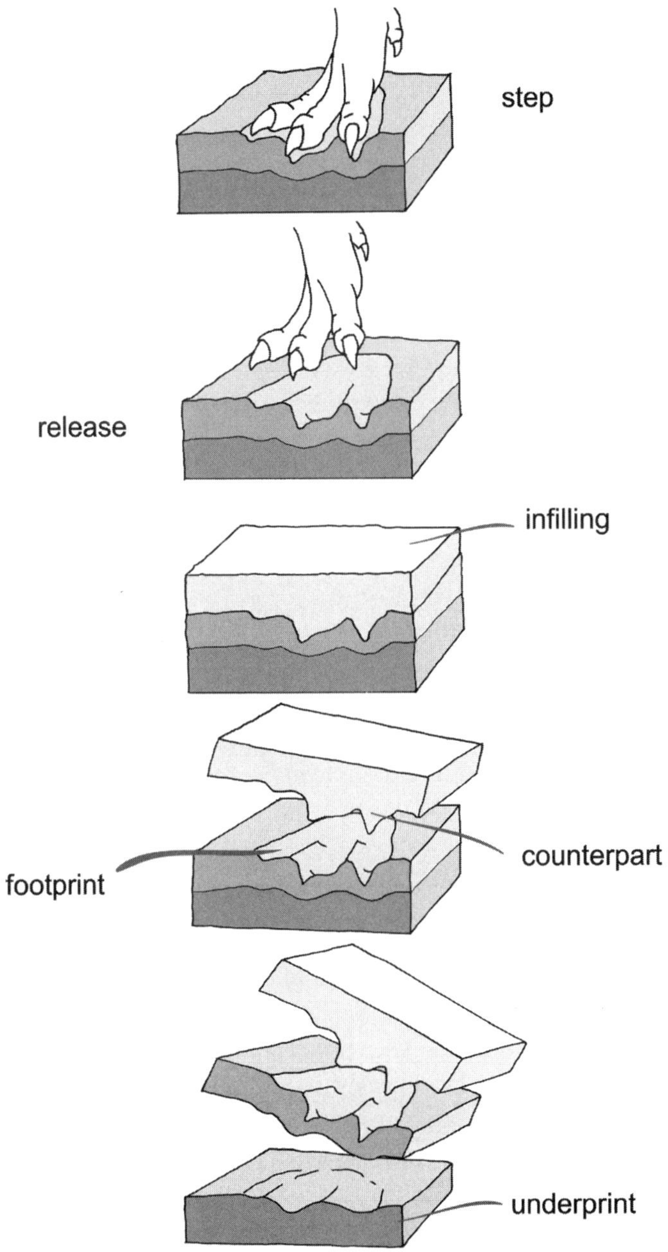


Footprints are formed when an animal treads on a deformable substratum. The impression of the foot may be preserved on the surface of the ground, and an "underprint," produced by the deformation of an underlying stratum, may be preserved as well. The hardening of the soil and its subsequent covering may, under particular conditions, lead to fossilization of the prints. *Drawing by F. Nodo.*

the environment. In fact, the footprints are without doubt autochthonous—that is, they were impressed where we find them today—whereas skeletal remains have often undergone more or less prolonged transportation, with the result that they can be found in a place very different from where the animal normally lived.

For this reason, research on fossil tracks has gradually improved over time. History is full of wrong interpretations, including some very curious ones. Just think, for instance, that up until the second half of the nineteenth century, the tridactyl footprints of theropods were thought to have been impressed in the rock by the weight of gigantic prehistoric birds; and so as not to go against the thinking of the period, some scientists even went so far as to attribute them to the raven that left Noah's ark after the Flood.

Today, ichnology—the science that studies fossil footprints—has made enormous progress, both because a large quantity of tracks have since come to light (according to Martin Lockley, there are more than fifteen hundred sites in the world, and that is only as far as dinosaurs are concerned) and because the use of more up-to-date study methods provides a truly impressive amount of information.

From a fossil footprint, expert ichnologists are able to deduce the size of the animal; its type of locomotion, speed, and posture; and even certain aspects of its social behavior. Dinosaur tail marks, represented by continuous grooves between the footprints, are extremely rare, which proves that these colossal animals walked with their tails held up off the ground. Parallel tracks left by several individuals all moving in the same direction uphold the hypothesis that different species of dinosaurs lived in groups. An example of this is footprints discovered by Roland T. Bird in 1940 at the site known as Davenport Ranch, in Texas, where at least twenty-three sauropods left tracks that indicate that they were moving in the same direction. Another site, discovered by Philip J. Currie in western Canada, preserves trackways of twelve ornithopods, similar to *Iguanodon,* that left unmistakable signs of having traveled in a herd (the equidistant tracks imply that the animals were traveling side by side).

Footprints of carnivores close to or overlapping those of herbivores can easily conjure up scenes of predation, as apparently took place in what is now Dinosaur Valley State Park (also in Texas), where the footprints of at least one carnivorous dinosaur following a group of sauropods can clearly be seen. The footprints discovered by Giuseppe Leonardi at Toro Toro in Bolivia are even more spectacular. It is easy to imagine that this spot was the scene of a true prehistoric drama in which at least thirty theropods hunted down a group of sauropods of gigantic dimensions.

By analyzing the footprints, it is even possible to deduce the approximate size of the animal that left them. In fact, ichnologists consider the length of the footprint to be about one-quarter or one-fifth (depending upon the type of dinosaur) of the length of the limb—that is to say, up to the hip. The English scientist R. McNeill Alexander developed equations that give an estimate of the animal's speed, starting from the length of its pace and presumed height at the hip. According to this formula, a dinosaur was walking if the length of its pace is less than four times the length of its foot, and it was running if the length of its pace is more. Even though a great majority of the known tracks were made by walking animals, analysis of trackways has shown that some dinosaurs were capable of moving very fast. Medium-sized carnivores going at full sprint reached speeds of around 30 to 40 kilometers an hour, while the herbivores, although larger and heavier, were much less agile. The performance of the dinosaurs of Lavini di Marco was rather poor (from 1 to 4 kilometers an hour). According to Giuseppe Leonardi, while they were undoubtedly slowed down by the muddy terrain, which hampered their movements, one has the strong impression that they behaved like the present-day caimans and crocodiles, which generally move very slowly, even though they are capable of darting forward at lightning speed when they are close to their prey. Another thing that can be said with certainty is that the Rovereto footprints denote solitary animals that did not travel in groups.

## Strolling along the Jurassic Beaches

So who was taking these peaceful strolls over the yellowish-white muddy stretches that bordered the Tridentine Sea 200 million years ago?

It is evident, of course, that a footprint changes in relation to the terrain on which it is impressed, the size and age of the animal making it, and even the different conditions that contributed to its preservation. All this makes it almost impossible to identify the exact species that left it. Nevertheless, a number of quite reliable conjectures can be made.

What strikes even the most inattentive spectator who goes to visit the Lavini di Marco is the enormous diversity of the fauna that left the telltale signs of their passage. Approximately 80 percent of the footprints are represented by distinctly separated three-toed prints, which are characteristic of the bipedal carnivorous dinosaurs (theropods). Fifteen percent of the tracks are composed of large oval footprints produced by the massive columnar feet of huge four-footed herbivores (sauropods). The final 5 percent consist of footprints with three short, juxtaposed, rounded-tipped toes that are attributed to smaller herbivores (primitive ornithopods).

The substantial numerical prevalence of theropod footprints does not necessarily mean that there were actually more carnivores than herbivores. The predator/prey ratio is always in favor of the latter, as is required by the laws that govern the trophic pyramid, but the predators had higher levels of activity because of a more efficient metabolism, which enabled them to explore vast hunting grounds. Therefore, within the same space of time, they produced more footprints than the herbivorous dinosaurs.

On the basis of skeletal remains, it is known that the theropods most widespread during the Lower Jurassic Period were the coelophysoids and the ceratosaurs. Consequently, it is among these that we should search to find the most probable authors of the tridactyl footprints of the Lavini di Marco. Coelophysoids and ceratosaurs were quite primitive theropods with four-fingered forelimbs. Many of them had developed bizarre crests or horns on top of their head. The jaws of the small- and medium-sized species were equipped with exceptionally long, sharp teeth, so their bite was their primary weapon, rather than the claws of their delicate forelimbs.

The largest of all the carnivorous dinosaurs that walked the Trento Platform were about as big as a dilophosaur (super-predators from 4 to 6 meters in length, weighing 500 to 800 kilos), while the smaller ones must have been similar to *Syntarsus,* a much more slender and agile animal, 2 to 3 meters in length and no heavier than around 100 kilos, whose bones have been found in South Africa. The former left impressive footprints on the rocks at Rovereto more than 30 centimeters long and lined up one behind the other, some of which are so well defined as to show the marks of their sharp talons.

Marco Avanzini, a geologist from the Museum of Trento, noted that within the same stratum, the footprints of the carnivores are often shallower and better defined (they were probably impressed on muddy surfaces less impregnated with water), while the tracks of the herbivorous dinosaurs have less detail and are more concentrated in areas where the sediment must have been covered with pools of stagnant, brackish water.

A fine tridactyl footprint of a theropod impressed in the gray limestone of Lavini di Marco. The impressions of the claws are also visible. © *Tridentine Natural Science Museum. Photo by C. Dal Sasso.*

These areas, several dozen meters wide, reoccur at various points of the site and seem to imply that the carnivores walked along muddy spits above sea level on the edges of the lagoons while awaiting their prey. The latter are represented by very primitive sauropods, whose footprints are unknown anywhere else in the world. Even though they were found in places quite far away from Italy (South Africa and India), the genera *Vulcanodon* and *Barapasaurus* are the only reference we have from which to picture the sauropods of the Lavini because they are the oldest dinosaurs of this type whose skeletons have been found. They were huge, lumbering beasts with a gigantic body (about 11 meters in length and weighing 3 or 4 tons), four column-like limbs, a long tail, and a long neck ending in a very small head. At Rovereto these animals left prints of their hind feet that are large, oval, and close together; each print measures up to 50 centimeters in length. The footprints of their forelegs, which are more widely spaced than the two hind feet, are small and horseshoe-shaped. (A fine example of these footprints is the track in the "Chemini Colatoio" labeled ROLM 1 and ROLM 11.)[3]

*Vulcanodon*-like sauropods existed in the Southern Hemisphere, which would appear to imply that the dinosaurs of Trento were immigrants from Gondwana.[4] More precisely, they are thought to have moved northward from southern Africa, and upon reaching the vast expanse of sea bordering the northern part of the continent, they pushed on to the furthermost tip, then represented by the Trento Platform. They could have arrived so far north only if there was a corridor that, at least sporadically, emerged above sea level.

Conversely, the paternity of the track labeled ROLM 9, which can also be seen in the "Chemini Colatoio," is still shrouded in mystery. The trackway is composed of twenty-two large footprints, each with three short, rounded, hoof-like toes, almost certainly made by the feet of a herbivorous dinosaur proceeding in an apparently bipedal gait. The footprints measure 35 to 40 centimeters in diameter and have extensive expulsion rims; that is, they are encircled by waves of mud pushed up by the animal as it trod over the ground, which indicate that the animal that left this trail must have been remarkably heavy, from 1 to 2 tons, and about 5 to 6 meters in length. According to Giuseppe Leonardi and Paolo Mietto, these footprints were probably left by a large ornithopod, an ancestor of the iguanodonts, whose footprints or bones, however, have never been found in rocks that old anywhere else in the world. Animals that could have produced footprints of this kind, such as the American

Lavini di Marco (Rovereto). The "Chemini Colatoio" with the track of a sauropod (ROLM 1) and, in the background, other strata outcropping at a lower altitude. © *Tridentine Natural Science Museum. Photo by C. Dal Sasso.*

Reconstruction of a probable hunting scene at Lavini di Marco, 190 million years ago. During the Lower Jurassic, the Trento Platform was inhabited by sauropod dinosaurs up to 12 meters in length, like this vulcanodontid, and by relatively primitive theropods (*left*). *Drawing by F. Fogliazza after a drawing by Gregory S. Paul.*

camptosaurs and the English callovosaurs, are known starting from the Middle-Upper Jurassic, at least 30 million years later. Another possibility is that the ROLM 9–type tracks were left by ornithischians similar to the scelidosaurs, dinosaurs provided with a light armor of dermal plates, which also lived in Europe in the Lower and Middle Jurassic, and which are considered to be the ancestors of *Stegosaurus* and *Ankylosaurus.*

Apart from those of the *colatoio* discovered by Chemini, trackways and isolated footprints belonging to other types of dinosaurs have been discovered over the years on the rock faces located on the uppermost part of the outcrop. Prints that are slightly smaller than those of the enigmatic ornithopod, more elongated due to the fact that the animal rested more heavily on the heel and had four toes instead of three, are thought to have been left by plateosaur-like prosauropod dinosaurs. Their appearance must have been truly awe-inspiring. They were huge beasts almost 8 meters long, which, thanks to their long necks, fed primarily on the leafy fronds of the tallest trees. They walked mainly on their hind limbs, but every so often they rested on their forelimbs, which were armed with powerful claws and a thumb with a huge, sharp scythe-like talon, which they may have used to defend themselves against predators and for grasping vegetation.

If confirmed, this fact would make the Lavini di Marco site the only one of its kind in the world, because it would prove the coexistence, in the same place and at the same time, of evolved forms and their ancestors: the prosauropods are in fact considered to be the ancestors of the sauropods—that is, of the same vulcanodonts. On the other hand, as we mentioned previously, in this geographic and temporal environment, the exception to the rule is the vulcanodont, whereas prosauropods are dinosaurs typical to the Upper Triassic and Lower Jurassic of Europe (numerous skeletons have been found in France, Switzerland, and Germany).

One of the largest outcropping layers at a higher altitude is literally strewn with footprints, to the point that it has been jokingly nicknamed the local dance hall. There are hundreds of footprints, unfortunately all rather shallow, and clearly visible only when the sun's rays are very low. Some measure a mere 3 centimeters and belong to tiny theropods, perhaps not much more than hatchlings. Others have weird shapes: *Anomoepus,* for example, is a pair of footprints side by side, about 20 centimeters long, each of which has three toemarks that continue rearward in an elongated groove. This track was left by a dinosaur "squatting" on its heels, producing two parallel grooves with the metatarsi of its feet.

The plateosaur is a prosauropod that lived in Europe at the end of the Triassic and beginning of the Jurassic. Dinosaurs similar to this one left numerous prints at Lavini di Marco. © *Natural History Museum of Milan. Photo by C. Dal Sasso.*

Footprints of this kind are extremely rare; only a few dozen exist in the world. Due to their shape and size, those discovered in Rovereto have been attributed to a small ornithischian, a very agile bipedal herbivore similar to *Fabrosaurus, Lesothosaurus,* or *Scutellosaurus.*

## The Enigma of Nourishment

One of the greatest enigmas that the experts are attempting to solve is how these enormous beasts could have survived in an area such as Lavini di Marco. For the carnivores the answer is easy and obvious, whereas for their prey the issue is rather more complex. The coasts and islands were undoubtedly covered with vegetation: dry-climate ferns, cycadeoids (plants vaguely resembling palms), and tall conifers related to the family of the Cheirolepidiaceae were a source of nourishment. But was it enough?

To understand the significance of the problem, a comparison can be made with the large herbivorous animals of our day and age, such as elephants: they need about 70 kilos of fodder a day and eat for almost ten hours at a stretch. The vegetarian dinosaurs must also have eaten a huge quantity of plants, and the vegetation growing on the small islands would probably not have been sufficient, even if the dinosaurs needed a lot less food than elephants. In other words, were they non-migratory, and the islands larger than we presumed, or were they migrating? And if so, where were they going?

At present, it is impossible to give a definite answer to this question. According to Giuseppe Leonardi's hypothesis, these dinosaurs were moving in search of food or on their way toward nesting areas. In fact, the footprints at Lavini di Marco have a preferential direction that goes from northeast to southwest and vice versa. This may also have been the same orientation of the mudflats that divided the stretches of water. The dinosaurs were obliged to follow them, in the only direction possible, to reach areas of dry land covered with vegetation, which lay some dozens and perhaps even hundreds of kilometers away.

Marco Avanzini has recently found confirmation of this hypothesis by analyzing the relationship between fossil pollen and footprints. The climate on the Trento Platform was rather arid, perhaps with little or no rainfall for long periods, even for decades. In the levels devoid of footprints, there are very few traces of pollen, and those that are found belong mainly to conifers with small, scaly leaves capable of withstanding drought. Conversely, in the rocks containing footprints, pollen is much more abundant and diversified, and there are signs of the temporary presence of fresh water. This would appear to prove that the dinosaurs arrived here when the climate became more humid.

On the basis of stratigraphic analyses, geologist have established that a complete climatic cycle lasted from 20,000 to 100,000 years (they refer to "megamonsoons," phenomena perhaps linked to the astronomical cycles), and that dinosaurs walked on the Trento Platform for some dozens of millions of years.

Since then, with the passing of time, the landscape has changed many times. Initially the entire area was submerged by the Tethys Sea (so called in honor of the ancient Greek sea goddess) and became a deep seabed. Enormous pressure, due to the accumulation of other sediments on top of the Jurassic sediments, compacted and cemented the calcareous mud of the carbonate plain, turning it into hard limestone rock. It later reemerged under powerful thrusts exerted on the sea floor as the tectonic plates shifted during continental drift. Immense upheavals slowly and gradually created the Alps. In the intervening eons, erosion has removed the layers bit by bit, peeling them away like the layers of an onion. And this is why the vestiges of the passage of these enormous prehistoric beasts are only now reemerging from the past to tell us about their world. In time, they are destined to disappear. Wind, rain, water, and ice will wipe them out forever. Before that happens, perhaps researchers will have the chance to obtain all the information they need to depict in all its finest details a scenario so unparalleled as to look like the brainchild of some very imaginative Hollywood movie director.

## A Little Bit of History

The footprints of Lavini di Marco are not the only dinosaur tracks found in Italy. The story of their discovery begins in 1941, at the height of the Second World War, the year German paleontologist Friederich von Huene published a description of a footprint discovered on Monte Pisano, near Agnano (Pisa). It consisted of a single tridactyl print only 6 to 7 centimeters long, which was called *Coelurosaurichnus* (now synonym of *Grallator*) *toscanus* and was attributed to a small-sized theropod dinosaur (probably a coelophysoid) measuring just over 1.5 meters from the snout to the tip of the tail and weighing just a few kilograms. The rocks in which it was discovered derive from sediments that were deposited in the delta of a river at the beginning of the Upper Triassic (Carnian, approximately 230 million years ago) and contain prints attributed to amphibians and non-dinosaurian reptiles. The Monte Pisano footprint, which is now housed in the Museum of Geology and Paleontology of the University of Florence, is consequently the first and oldest sign of the presence of dinosaurs in Italy.

Many years had to pass before other dinosaur footprints were discovered in Italy. Then, in 1985, Paolo Mietto, a professor at the University of Padua, reported the presence of tracks on an enormous boulder discovered by Vittorino Cazzetta in Cadore at the foot of Monte Pelmetto, on the eastern side of the Dolomites in the province of Belluno. Cazzetta, who took a keen interest in paleontology and early history, had also made some remarkable discoveries in archeology. In fact, it is thanks to him that the Mesolithic burial site of the Mondeval man was found at an altitude of 2,150 meters, complete with the skeleton of a hunter from 7,500 years ago together with his funerary objects. Vittorino Cazzetta's life ended tragically when, at the age of 49, he fell into a chasm in an inaccessible area in the Dolomites. He had set out, as he often did, on a number of explorations in the mountains on August 10, 1996, but he never returned home. The search went on for months, but to no avail. A year later his body was found, huddled in a deep ravine into which he had fallen. In all probability, due to the fractures he suffered during the fall, he was unable to climb up again and died while waiting to be rescued.

At Monte Pelmetto, the surface of the boulder with the prints, which is oriented vertically, contains no fewer than five dinosaur trackways. Three of them have footprints 6 to 7 centimeters long impressed by small predatory theropods, while a fourth track with larger footprints (10 to 12 centimeters) has been attributed to a primitive type of bipedal ornithischian. Last, the footprints of the largest animal (the marks left by the foot are approximately 15 centimeters long) could be those of a prosauropod dinosaur no more than 2 to 3 meters in length and similar in appearance to the plateosaur but smaller and more slender.

The huge boulder forms part of the accumulation of a landslide that broke away from the western edge of Monte Pelmetto. Searches in the underlying scree, thanks also to the activity of the Val Fiorentina Museum (Selva di Cadore), have recently led to the recovery of smaller blocks of

The Monte Pelmetto boulder. The track oriented diagonally toward the top is attributed to a primitive ornithischian; the track at the bottom, on the right, consisting of larger prints, was made by a prosauropod. *© Tridentine Natural Science Museum. Photo by G. Teruzzi.*

rock containing numerous other footprints belonging to dinosaurs and "thecodonts" (a group of primitive archosaurs that includes the probable progenitors of the dinosaurs).

Another site with reptilian footprints dating back to the Upper Triassic was identified in the area around Lerici (La Spezia) in 1987, thanks to the watchful eye of a 12-year-old boy, Ilario Sirigu. After news of the find spread, a number of prints were stolen, but they were recovered after an anonymous tip led police to a bag left near the cemetery in Lerici. There are about 100 footprints impressed on a rock surface close to the sea, which date to approximately 225 million years ago. So far, experts have recognized three species of dinosaurs and two species of "thecodonts."

Among the dinosaurs that trod this area, which during the Triassic must have been a bleak and arid coastal plain crossed by seasonal streams, types similar to *Coelophysis* and *Fabrosaurus* have been identified. The former were long-limbed carnivorous dinosaurs approximately 2.5 meters in length belonging to the infraorder of the Coelophysoidea, which left tridactyl prints with slender toemarks and small nails on the terrain. The latter were herbivorous and belonged to the order Ornithischia. There remains some doubt as to the interpretation of several footprints that appear to have been left by four-toed feet with large talon-like claws. They could belong to a prosauropod, perhaps similar to the genus *Sellosaurus,* a herbivorous dinosaur about 3 meters long. Among the "thecodonts" there were undoubtedly aetosaurs, quadrupedal herbivores measuring over 2 meters and provided with sturdy armament to protect them from predators, and rauisuchians, fearsome carnivores that could reach even larger dimensions.

Two large three-toed footprints (approximately 30 centimeters in length), which are named *Eubrontes* and can be attributed to theropod dinosaurs of huge dimensions, were also discovered on the eastern side of the Dolomites in the area of Belluno, on a boulder on the lower slopes of the Tre Cime di Lavaredo. Other findings were later reported from the western side of the Dolomites. In fact, in 1994 Giuseppe Leonardi identified the presence of a tridactyl footprint near Cima Puez, in the province of Bolzano. All the Dolomite footprints were impressed in the calcareous ooze of a vast tidal flat that extended over northern Italy until the end of the Triassic Period, about 215 million years ago. The prolonged sedimentation across this tidal flat contributed to the genesis of a well-known geological formation, the Dolomia Principale.

Between 1994 and 2000, a dozen or so boulders of Dolomia Principale with other footprints of large reptiles were discovered in the Parco delle Dolomiti Friulane between the villages of Andreis, Claut, and Cimolais in the foothills of the Carnian Alps (Pordenone province). The first clue, the footprint of a large carnivorous dinosaur, was spotted by a young schoolteacher when he stopped at the foot of a waterfall to fill a flask with water while accompanying his students on a school outing. However, the person who discovered most of the prints, which were located in isolated gorges and on desolate peaks, was Mauro Caldana, a nurse whose hobby was ornithology and who rambled through the mountains on bird-watching expeditions. Although the prints most frequently found are medium-sized tridactyl footprints (between 15 and 25 centimeters in length) attributable to theropod dinosaurs, in these boulders there are also footprints that measure 35 centimeters in length. According to the paleontologist Fabio Dalla Vecchia, who studied the prints with Paolo Mietto, these *Eubrontes*-like tracks are particularly important because they belong to enormous carnivores up to 7 meters long, which left no skeletal remains as tangible proof of their existence in Italy during the Triassic Period. Together with those of the theropods, there are also traces of prosauropods of various dimensions and thecodontian reptiles. Unfortunately, one of the fossil-bearing boulders was destroyed, and with it a long bipedal dinosaur trackway.

In addition to the important site at Lavini di Marco, numerous Jurassic localities with dinosaur footprints have been discovered in Trentino and in the western Veneto area. Guido Roghi, a paleobotanist from the University of Padua, whose work consists of dating rocks and reconstructing the vegetation of the past by means of fossil pollen, discovered the Bella Lasta site in the Veronese Upper Lessini (Revolto Valley), which contained tridactyl footprints of theropods of huge dimensions (*Kayentapus*). We owe to Marco Avanzini of the Tridentine Museum of Natural Sciences the reporting of other new outcrops (Monte Pasubio, northern Lessini, Becco di Filadonna, Valle del Sarca, Marocche di Dro) with footprints attributed to variously sized theropods and to herbivorous dinosaurs, for the most part sauropods, while we owe the discovery of the Chizzola site to Giuseppe Leonardi.

However, until the summer of 1994, there had never been any evidence of dinosaur tracks dating back to the Cretaceous period. Such prints were found by Sandro Venturini, a geologist from Friuli employed by an Italian oil company, in a place where no one would ever have gone to look for them: on the piers at Ravenna. Venturini was on the Porto Corsini pier with his wife and daughter during a day's vacation at the seaside when all of a sudden, before his very eyes, as clear as clear could be, there appeared a large tridactyl footprint in relief. The print was a counterpart, the cast of the original 35-centimeter-long footprint, but it was proof that a large theropod dinosaur had walked over the ancient mud that, turned into rock, had been incorporated into a breakwater. Subsequent investigations enabled experts to establish that the block, which is now housed in the Museum of Natural Sciences in Faenza, dates back to the Lower Cretaceous and came from a quarry located on the lower slopes of the Altopiano del Cansiglio, in the province of Pordenone. A study of the rock, carried out by Fabio Dalla Vecchia (who together with Sandro Venturini described hundreds of dinosaur footprints in the Cretaceous system of rocks in Istria), led to the further discovery of a classical horseshoe-shaped print, 30 centimeters wide, produced by the hand of a sauropod dinosaur. Thus, carnivorous and herbivorous dinosaurs lived not far from what is now, some 120 million years later, the popular ski resort of Piancavallo and the American military base of Aviano.

## Serendipity in Apulia

The most extraordinary and unexpected footprint discoveries have recently been made in Apulia. In the spring of 1999, the geologists Massimo Sarti and Michele Claps of the Marine Sciences Institute of the University of Ancona were in the heart of the Altopiano delle Murge prospecting for hydrocarbon deposits. One afternoon, as they were exploring the countryside around Altamura, a small town of 63,000 inhabitants in the province of Bari, Sarti and Claps stopped at the entrance of a large quarry from which coarse aggregate is extracted for the building industry. The gate was open, and after some hesitation they decided to go in. Almost immediately they realized that something was rather odd. The sun was low in the sky and cast shadows on the ground before them, highlighting hundreds of holes of different dimensions. They were standing before what is undoubtedly the largest tracksite in Italy, and perhaps one of the largest in all of Europe.

The incredible fact, apart from the extremely large number of footprints, is that the fossil-bearing stratum had already been freed from the overlying rock, as though it had been worked on by paleontologists who

Partial view of the De Lucia quarry of Altamura. The hundreds of little holes that can be indistinctly seen on the limestone pavement are all dinosaur footprints. *© Italian Ministry for Cultural Heritage and Activities, Archeological Superintendence of Apulia. Photo by C. Dal Sasso.*

intended to bring the footprints to light, and that this operation had been carried out for years wholly involuntarily. In fact, the quarry workers, having come across a particularly hard and massive layer of calcareous rock, had perhaps considered it not worth extracting and had decided not to remove it but instead to use it as a surface over which to drive their vehicles. But none of them had ever wondered what those strange holes might be.

Despite the fact that the discovery is relatively recent and the studies on it have only just begun, according to a preliminary estimate there are approximately 30,000 footprints dating back to 75–80 million years ago, impressed on a calcareous area of 12,000 square meters. This is an impressive number that could quite easily be doubled by merely widening the quarry by a few dozen meters. In fact, the Calcari di Altamura lie in a perfectly horizontal position and are not, or are only very slightly, dislocated by faults. This means that the bed containing the dinosaur tracks can be followed in all directions for hundreds of meters. And to

Altamura (Bari). Detail of a long track attributed to a herbivorous dinosaur that moved with a quadrupedal gait. © *Italian Ministry for Cultural Heritage and Activities, Archeological Superintendence of Apulia. Photo by C. Dal Sasso.*

think that, according to the indications on the geological maps, in the Upper Cretaceous period Apulia was completely submerged by the sea!

On observing the trackways (one of the most complete of which consists of no fewer than 176 footprints in succession), one thing that is immediately evident is that they do not have a preferred orientation. They cross one another, change direction, and overlap. The impression is that these are not animals heading toward a certain place—for example, toward an island by following a sandbar left high and dry by the tide—nor are they migrating herds. Perhaps this was a grazing area. It is not by mere chance that the footprints found so far at Altamura all belong to herbivorous dinosaurs.

The first hypotheses suggest that the tracks were made by hadrosaurs and sauropods. The former are well documented by long strides made by feet with three not-very-long toes (sometimes with traces of the heels) and hands with four fingers joined by a fleshy horseshoe-shaped pad, and they must have been relatively small (it would appear that they were no more than 4 to 5 meters in length). The latter left the classic oval footprints, which were also medium to small (25 to 30 centimeters in diameter) in comparison to those left by standard-sized sauropods. Perhaps, as indicated by footprints wider than they are long with four short, rounded toes, there are also traces of huge armored beasts, such as nodosaurs and ankylosaurs. These herbivorous dinosaurs belong to the suborder of the Thyreophora and were covered with a mosaic of dermal plates and spines. They are known from skeletons found in North America, Asia, and Europe, but their footprints are extremely rare.

There are consequently a great variety of shapes and clearly defined details, made possible by the very fine grain of the sediment, which makes the tracksite of Altamura spectacular and fascinating. In several footprints, even folds of skin appear to have been impressed.

Strangely enough, these footprints all lack the characteristic expulsion rims, those raised borders created by the weight of the animal shifting the soft mud. Perhaps this is because, as explained by Umberto Nicosia of La Sapienza University of Rome, who is in charge of studying the site, within the thickness of the layer the grains of sand were spaced apart from one another by microscopic voids occupied by cyanobacteria and algae, which became compacted only when the dinosaurs trod on them, giving the sediment a plastic behavior. Moreover, this mucilage would have prevented the sediment (which is actually carbonate ooze and consequently devoid of clay) from seeping back into the footprints after the

One possible author of the broad four-fingered prints impressed in the limestone of Altamura (*top*) is *Struthiosaurus* (*bottom*). Skeletal remains of this genus of ankylosaur (a small-sized nodosaurid) have been found in Austria, France, Romania, and Hungary. *© Italian Ministry for Cultural Heritage and Activities, Archeological Superintendence of Apulia. Photo by C. Dal Sasso. Drawing by F. Fogliazza.*

passage of the animals by acting as a natural supporting sealant. The inner edge of the deepest footprints at Altamura is a vertical wall, and no non-argillaceous sand or mud can withstand an inclination of 90 degrees on its own.

In addition to the lack of expulsion rims, another difference in comparison to the Rovereto tracksite is that in the Calcari di Altamura there are no signs of those thin laminae that indicate frequent interruptions in sedimentation. The layer with the footprints and the upper layer seem to have been deposited within a very short space of time, as though they were caused by two flood tides.

The Apulian footprints also tell us that these enormous prehistoric beasts must have frequented the area within a very short period of time, perhaps even simultaneously. In fact, when the tracks cross one another, it can clearly be seen that the footprints of the track made first have been plastically deformed, and not crushed or cracked, by the passage of the second animal: this means that the mud was still fresh.

However, as in the case of the Lavini di Marco, the biggest question concerns the dinosaurs' nourishment. If these animals were non-migratory, where did they find the enormous quantity of plants they needed to survive? And if they were migrating, where were they going? In proportion to their size, they would have needed vast extensions of dry land with forests and rivers at their disposal, a landscape totally different from the arid and semi-barren carbonate plain that this area is presumed to have been. Under ecological conditions such as these, plant productivity must have been rather low, and certainly insufficient to feed hundreds of dinosaurs, each of which would have needed about 100 kilos of "fodder" a week.

With respect to their origin, there are not many possibilities. The paleogeographic models most widely accepted to date hypothesize that the Apulian Plate (corresponding to the present-day Apulia) was surrounded by deep straits from as early as the Lower Jurassic. At the present stage of knowledge, according to Umberto Nicosia, the only possible route linking it to other areas of dry land would have been toward the east, by means of a series of microplatforms situated along the area that probably joined the present-day geographical regions of Apulia, Dalmatia, Greece, Bulgaria, and perhaps also Turkey.

A story somewhat similar to that of Altamura could be told about the footprints discovered on the Gargano Promontory near the village of San Giovanni Rotondo in June 2000. In fact, here, too, the protag-

A handprint and a footprint belonging to the track of a medium-sized ornithopod, probably a hadrosaur. Note the lack of expulsion rims in the calcareous sludge. Pen for scale. © *Italian Ministry for Cultural Heritage and Activities, Archeological Superintendence of Apulia. Photo by C. Dal Sasso.*

Reconstruction of one of the possible authors of the footprints of Altamura: a hadrosaur of the subfamily Hadrosaurinae, of which European species are known, all lacking nasal, frontal, and nucal crests. *Drawing by F. Fogliazza.*

onists are geologists, this time from the University of Ferrara, and the setting is a quarry. During one of their routine trips to study the geology of the area, Alfonso Bosellini, Piero Gianolla, and Michele Morsilli found themselves face to face with a number of large boulders containing dozens of clearly visible tridactyl footprints (actually counterparts of the imprints), from 15 to 40 centimeters long, for the most part still arranged so as to form trackways. Most of them belong to large bipedal dinosaurs, both carnivorous (theropods) and herbivorous (ornithopods).

The Gargano footprints are much older than those of Altamura and date back to approximately 120 million years ago (Lower Cretaceous). The discovery was unexpected precisely because of the age of the rocks, which corresponds to a period during which, according to paleogeographic reconstructions, the Apulian Plate should have been isolated in the middle of the eastern Tethys Sea. There are two possible explanations: either the Gargano dinosaurs were already present before the separation of the platform from the African continent and evolved under isolated conditions, or they freely crossed over by means of natural bridges formed by strips of the platform raised above sea level, which were presumed not to have existed.

The discovery of these tracks was followed in the fall of 2000 by the finding of footprints on the pier of Mattinata, also along the Gargano coast. About a dozen of them (the largest measure just under 30 centimeters) were impressed on two enormous blocks of limestone that formed part of the barrier erected to defend the harbor. According to the experts, these footprints were also left on the shoreline of extensive tidal flats, which were mainly composed of the accumulation of carbonate ooze and fine algal carpets. The blocks of fossilized sediment might have been quarried from the Late Jurassic Sannicandro Limestone Formation, which outcrops only in the Gargano area. They contain elongate tetradactyl footprints of an unusual theropod (*Theroplantigrada* sp.), medium-sized theropod footprints (?*Therangospodus* sp.), isolated footprints (*Megalosauripus* sp.), and a few marks that are difficult to attribute. This theropod-dominated association represents a third dinosaur-rich level in the Apulian Platform; together with the findings mentioned above, it demonstrates that part of southern Italy was inhabited by large dinosaurs for quite a long time.

# 3

# *SCIPIONYX*—"CIRO" TO HIS FRIENDS

The Natural History Museum in Milan, with its neo-Romantic–style building, has little in common with the science-fiction laboratories of *Jurassic Park.* And yet, in that imposing building in the heart of the capital of the region of Lombardy in northern Italy, they reconstruct dinosaurs. In October 1997, down in the basement of the museum, behind the scenes and out of sight of visitors, the first Italian dinosaur was being resurrected. A little model that started almost as a joke was a few months later to become the emblem of discovery. It would soon appear in the newspapers and on television throughout the world.

At first glance, *Scipionyx samniticus* (the scientific name given to him by the researchers who studied him) may seem a little disappointing. He is by no means huge and terrifying, as most people imagine dinosaurs to be. His little fossilized skeleton is impressed on a slab of limestone not much bigger than a shoe box. And yet, despite his tiny size, in a very short time he has become a real celebrity, even earning himself the cover of the prestigious scientific journal *Nature.*[1] As word of his discovery spread, so did his commercial value, which is currently estimated at around 2.5 million euro.[2] But why are paleontologists throughout the world showing so much interest in the little Italian dinosaur?

Apart from the "dinomania" sparked by *Jurassic Park,* which over the past few years has considerably whetted public interest in paleontology, the worldwide attention paid to *Scipionyx* is of a purely scientific nature. The reasons for this interest are manifold:

- Of all the dinosaur skeletons discovered to date, it is one of the most complete.

- It is an extremely rare immature specimen, a newly hatched "baby."
- It is one of the very smallest dinosaurs in the world.
- It possesses an original combination of anatomical features.
- It bears witness to the fact that during the Mesozoic Era, the Italian peninsula was not completely submerged by water.
- It is the first dinosaur in the world in which well-preserved internal organs can be seen.

To get a clearer idea of how exceptional it is, consider that although nearly nine hundred species of dinosaurs are known to date, more than 85 percent of them were established on the basis of skeletal fragments: one or two bones of the limbs, a few vertebrae, or, in the more fortunate cases, part of the skull. More often than not, we have to make do with a fragment of the jawbone or a single tooth. A great majority of the skeletons set up and exhibited in museums consequently consist of copies of the few existing original skeletons, which, however, are very often composite—that is, they were assembled from the bones of different individuals.

## An All-Italian Discovery

It all started one warm spring day in 1981. Giovanni Todesco, a footwear technician from Verona, was in Avellino on business. His passion for fossils incited him to go on excursions on weekends around the mountains in the Campania region, often accompanied by his wife and children. That day, their destination was Pietraroia, a village with 710 inhabitants, located at an altitude of 960 meters on the Matese massif, in the province of Benevento.[3] At that time a nearby limestone quarry, which had been exploited extensively in the past to obtain stone for buildings in the area, had practically been abandoned and turned into a refuse dump. Among the garbage and the mechanical scrapers that were removing material to construct a roadbed, Todesco pulled out some slabs from the cut in the stratified rock on which fragments of fish were fossilized. Then, his curiosity aroused by the thickness of several bones emerging from the surface of another small slab, he carefully removed some fragments that appeared to match with the first slab and that contained other parts of the same animal. While reassembling the slabs, he caught a glimpse of the skeleton of a small animal with limbs just a few millimeters beneath the surface of the sedimentary rock. It was certainly

The skeleton of the dinosaur of Pietraroia with the bones in perfect anatomical connection. The only parts missing are the ends of the tail and of the hind limbs. It measures just a little over 23 centimeters. The limestone slab in which it is fossilized dates back to approximately 110 million years ago. *© Italian Ministry for Cultural Heritage and Activities, Archeological Superintendence of Salerno, Avellino, and Benevento. Photo by L. Vitola. Reprinted by permission from* Nature *392, no. 6674. © 1998 Macmillan Publishers Ltd.*

not a fish, and this really aroused his curiosity. Todesco knew that that place was historically renowned as a site containing fossilized fishes dating back to the Cretaceous Period.[4]

After returning home to San Giovanni Ilarione (Verona), Todesco tried to clean the strange fossil with engraving tools and iron nails. The outline that emerged reminded him of the famous *Archaeopteryx* (the oldest known bird, considered to be the connecting link with the reptiles), but the discoverer's investigations went no further, and the "doggy," as he nicknamed it because of its sharp teeth, ended up in a drawer at home. There it remained until spring 1992, when Todesco met Giorgio Teruzzi, a paleontologist who worked at the Natural History Museum in Milan. Todesco told Teruzzi about his puzzling find and asked him to take a look at the fossil.

Twelve years after its discovery, the intriguing specimen lay on a table under the light of a lamp. A compressed little skeleton with reddish-brown bones appeared distinctly on the surface of a yellowish-gray limestone slab measuring 38 by 26 centimeters. Incredulous, shaken by emotion, the Milanese paleontologist held his breath amid the general hush, then whispered: "I am the first Italian paleontologist to see the first Italian dinosaur."

Even though he was a paleontologist, Teruzzi was not an expert on reptiles, and he immediately sought confirmation. He picked up the telephone and dialed the home number of a young colleague. "I'm here near Verona to see that strange item," he told me. "It's a reptile . . . it's very tiny. . . . But it has three clawed fingers on the forelimb." That night was a sleepless one.[5]

Over the following weeks, the Milanese researcher (who, with a touch of exaggerated local pride, had promptly rechristened the little reptile with the nickname Ambrogio—the patron saint of Milan) sought and found a second confirmation from Giuseppe Leonardi, a priest of the Congregation of the Schools of Charity (Cavanis Institute) and a paleontologist of great renown. Until a few years ago, there were no paleontologists in Italy who specialized in studying dinosaurs. No one had ever found a dinosaur skeleton in the territory, and no one had ever needed to study them. But Father Leonardi was a great specialist in dinosaur tracks. An eccentric personality with an almost legendary charisma, a lover of geological sciences by family tradition, he had worked in Latin America for many years as a missionary but also as a researcher, paleontologist, and explorer. Since returning to Italy, Leonardi was busy

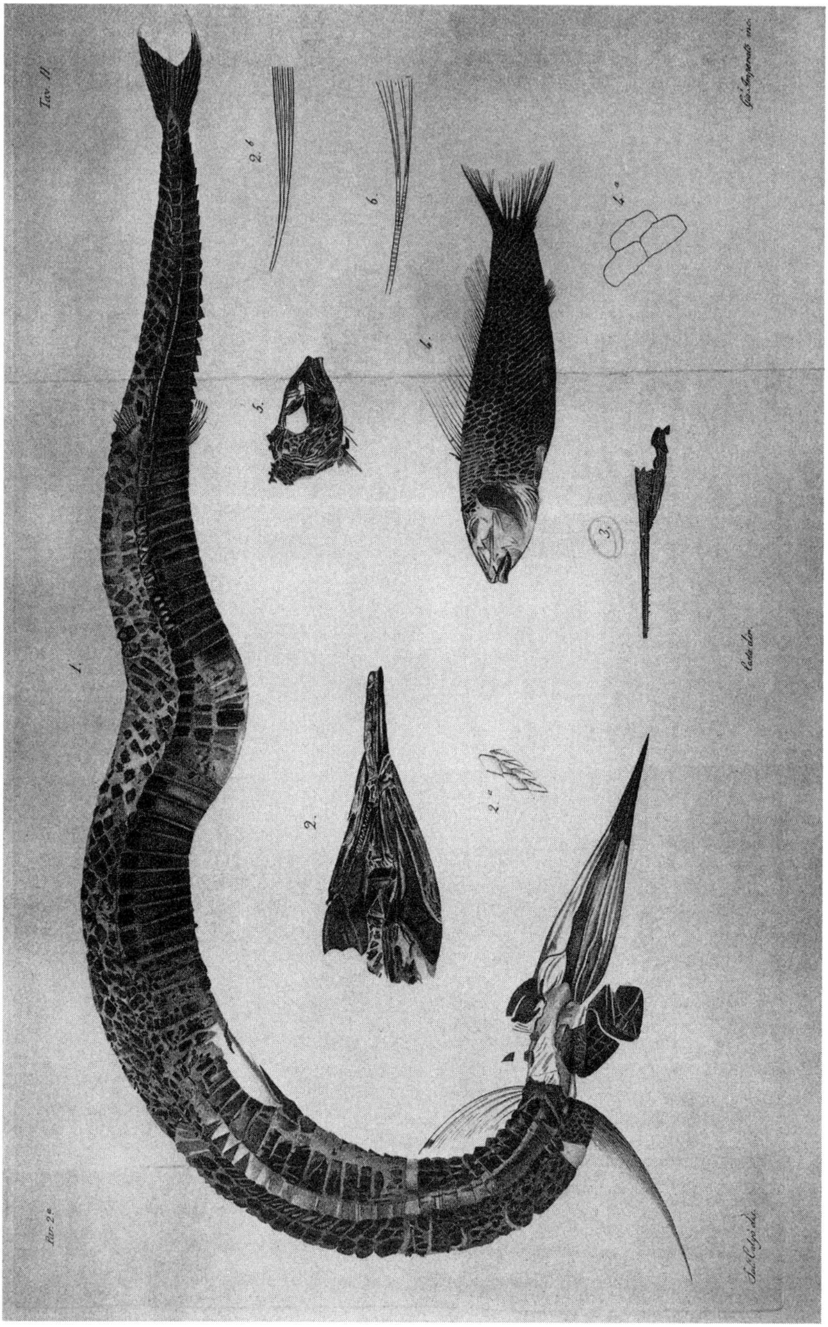

Plate from O. G. Costa (1854–1856) depicting a large *Belonostomus crassirostris* and other fossil fishes discovered at that time in the Cretaceous limestone of Pietraroia. *Photo by L. Spezia.*

studying the dinosaur footprints of the Lavini di Marco, near Rovereto, which had recently been discovered.

Due to a curious circumstance, the dinosaur "appeared" to Father Leonardi in a church in Corsico, on the outskirts of Milan, where he had gone to receive Cardinal Martini on a pastoral visit. Just as the preparations for the ceremony were in full swing, Teruzzi arrived with some slides taken at Todesco's home. Father Leonardi, eager to see them, put them in the projector that had been set up in the church for a catechism lesson. Just before the archbishop arrived, he had them projected onto the wall above the altar, to the great bewilderment of those present.

One of the most pressing problems related to the discovery of *Scipionyx* was how to help the fossil come out of hiding. In Italy, fossils are the property of the state, and Todesco, having one of such great importance, risked big problems with the law. The only solution, for the good of the dinosaur and its discoverer, was for the precious find to be handed over to a public institution.[6]

This situation was remedied thanks to the intervention of Franco Capone, a scientific journalist who for a time had been the press agent of the Natural History Museum in Milan. Through Capone, news of the existence of an Italian dinosaur landed on the city desk of the weekly news magazine *Oggi,* which immediately got wind of the discovery. There then ensued a delicate mediation with Todesco, who, eventually convinced of the public importance of the discovery, decided to hand the dinosaur over to the competent authorities. *Oggi* would publish the story of the discovery and accompany Todesco on his journey to return the precious specimen to its region of origin.

On October 15, 1993, Giovanni Todesco crossed the threshold of the Archaeological Superintendence of Naples clutching an overnight bag. After the customary exchange of greetings and the astonishment of the officials at the announcement of the object to be handed over, Giovanni was asked where the important specimen was. At this point, all that was left to do was open the overnight bag. With some regret, Todesco left the fossil at the Superintendence in Naples, which would later hand it over to its branch in Salerno, which had territorial responsibility for safeguarding the fossil site of Pietraroia.

The little Italian dinosaur was about to become a star, but it still didn't have a scientific name. The editorial staff of *Oggi* thought it would be a good idea to prepare a life-size reconstruction of their star (a first model was made by Silvo Tassinari and Lorenzo Baides) and to give the dinosaur a "made in Italy" name that was also rather cute. The journal's editor-in-chief, Pino Aprile, then dubbed the diminutive beast "Ciro," a typically Neapolitan name.

Around the same time, the paleontology review *Paleocronache* published the first scientific report on the dinosaur, written by Leonardi and Teruzzi, including a preliminary description of the specimen. The daily newspapers acknowledged the discovery with only a few lines. Apart from a popular scientific article or two, until 1998 the first Italian dinosaur was no longer discussed. Despite its nickname, "Ciro" did not have a clear identity. Because the specimen had still not been studied in detail, its discovery could not be fully exploited. In fact, it was not possible to study it until its entire skeleton had been exposed, and that would take a long time.

## Autopsy of a Dinosaur

At the beginning of 1994, a long and painstaking operation was underway in top secrecy at the Restoration Laboratory of the Archeological Superintendence in Salerno. Despite its small size (or rather, precisely for this reason), "Ciro" needed to undergo careful preparation to remove the sediment that still covered it. It was estimated that this operation would take a couple of weeks. However, due to the uniqueness of the fossil and the unexpected discovery during the work of part of the internal organs and other very delicate soft tissues, it became necessary to proceed with the utmost caution. In the end, preparation of the fossil took many months.[7]

Paleontological preparation is an extremely precise and drawn-out process. On small, fragile specimens, the operation is performed with the aid of a microscope and must be done by hand, using micro-chisels and increasingly finer steel needles with which the sedimentary rock can be chipped away a grain at a time. It takes a steady and sensitive hand, but above all lots of patience. "Ciro," who was embedded in a thin calcareous slab, could not be brought out in full relief, so the work consisted in creating a bas-relief, leaving the rock as a base to support the bones.

The preliminary and roughest part of the preparation had been carried out by the discoverer by hand and without a microscope. Todesco had reassembled the calcareous layer containing the fossil and had then revealed most of the skeleton. Since he did not find the tail, he reconstructed it using synthetic resin. He had also covered the exposed bones

The author in 1994 during the preparation of the Pietraroia dinosaur. *Photo by G. Teruzzi.*

with sticky vinyl glue, which had protected the fossil from further deterioration but was now hindering the cleanup and preventing the fossil from being clearly visible. During the task of removing the false tail, the glue, the sedimentary rock, and the fragile bones had to be consolidated with special impregnating resins as they gradually emerged. This task required eight to nine hours a day with eyes glued to the microscope at maximum magnification. And yet the time flew by in the twinkling of an eye, and each new day brought surprises. Beneath the glue and the grains of sediment there emerged not only the usual bare bones, belonging to a previously unknown species, but also muscle fibers, the cartilaginous rings of the trachea, the keratin of the nails, the red patch of the liver, and the intestinal folds.

Nobody had previously performed such an autopsy on a dinosaur. To the eyes of the person looking into the beam of light of a stereomicroscope, even the slightest scratch looks like a deep valley. Traveling along that tiny body, which measured a mere 23 centimeters, was like soaring over a mountain range. But it was not a familiar landscape. It was more like a journey into space, to the outermost boundaries of the known world, not over vast distances but under the microscope. In the crowded center of a city, 5 meters from the hustle and bustle of the stores, in the silence of a small room, under the microscope lay a whole universe to be discovered in just a few square centimeters.

## Scipio's Talon

Studying "Ciro" required even more time: about a year and a half. After the preparation, the Archaeological Superintendence of Salerno, Avellino, and Benevento, led by Giuliana Tocco Sciarelli, granted the necessary authorization to study the dinosaur. In order to compare the Italian dinosaur with already known species, it was necessary to read all the books and articles published on the subject throughout the world, and in order to examine still-unpublished specimens, it was necessary to visit collections in several American museums. Once it was ascertained without a shadow of a doubt that "Ciro" belonged to an unknown species, we decided to aim high: the formal description—the article containing the diagnosis of the new species—would be sent to the editor of the English journal *Nature*.

But at the end of the research, the dinosaur still did not have a scientific name. None of the many names suggested (*Italoraptor, Italosau-*

Preparation of the most delicate fossils, like the little *Scipionyx,* is a time-consuming operation that must be carried out by hand under a microscope with the aid of metal points as sharp as needles. *© Italian Ministry for Cultural Heritage and Activities, Archeological Superintendence of Salerno, Avellino, and Benevento. Photo by C. Dal Sasso.*

*rus, Microraptor*) were really satisfying.[8] What was needed was an incisive, original, and perhaps historical name, in keeping with the Italian culture. One day while talking about Scipione Breislak, the scholar who no less than two hundred years previously had been the first to describe the fossils of Pietraroia, we hit upon the compound name Scipio-onyx (*onyx* in Greek means claw). The name also recalled Scipio Africanus, the general in the Roman army who defeated Hannibal: "Ciro" had powerful curved claws and, like the Roman general, had lived and "fought" in the Mediterranean area, or rather, in that region that the Latins called Samnium. *Scipionyx samniticus* was a perfect name.[9]

Before being accepted, and with the condition that no news of it leak out until the date of publication, the article remained in the *Nature* files for almost six months. But the delay was due precisely to the exceptional nature of the discovery. Despite the copious high-quality photographic documentation (produced by Leonardo Vitola of the Superintendence of Salerno) accompanying the request for publication, the editor and scientific consultants of *Nature,* perhaps fearing a hoax, chose to proceed with caution. After all, the article reporting this most important discovery had been submitted to the world's most authoritative scientific journal by two illustrious nobodies.

The months rolled by. Then on March 20, 1998, without notice, the Natural History Museum of Milan received via fax a press release that was shortly to be released to the international news agencies. In it the editorial department of *Nature* listed, in order of importance, the four leading articles that the journal intended to publish the following week. At the top of the first page appeared a title in underscored capital letters: "RELICS: SCIPIO RIDES AGAIN" (COVER).[10]

This announcement was the culmination of a long and arduous operation conducted amid hundreds of difficulties by an all-Italian research team. Italy's debut into the world of dinosaurs couldn't have been better. The specimen was acclaimed by the media throughout the world as one of the most important fossil vertebrates that had ever been discovered. And a little village, practically unknown to the world, suddenly became a universal capital of paleontology thanks to the fossil of a little reptile embedded in a slab of limestone no bigger than a pizza: a sign from the past that nobody would have imagined could be so small, and yet so great.[11]

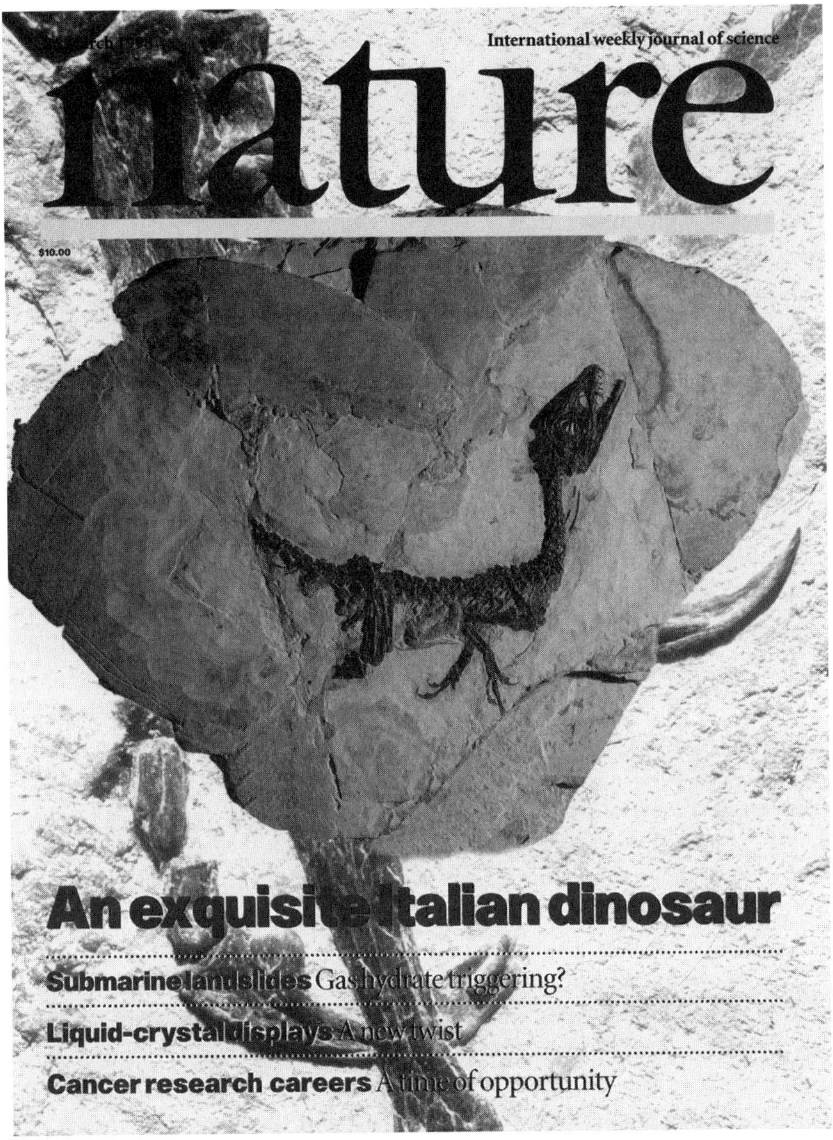


The front cover dedicated by *Nature* to the scientific christening of *Scipionyx samniticus. Reprinted by permission from* Nature *392, no. 6674. © 1998 Macmillan Publishers Ltd.*

## Every Inch a Baby

*Scipionyx samniticus* is missing most of its tail and the ends of its hind limbs, but the rest of the skeleton is complete and perfectly articulated. Each skull bone is where you would expect to find it. Each vertebra, each phalanx of the hand is in anatomical connection with the others. Thus it is highly unlikely that the little dinosaur suffered damage from predation or transport while alive or shortly after death. The animal must have been fossilized intact and the missing part lost at the time of its collection. How, where, and why the little dinosaur died, we have no way of knowing. The body lies on its left side, and the head is upturned with the mouth open. This position of fossilization is not an indication of suffocation but simply the result of postmortem contraction of the neck and jaw muscles, a common phenomenon that occurs regardless of the cause of death and the type of environment in which it occurs.

Apart from the general shape of the body, from which it can be perceived that *Scipionyx* is a carnivorous dinosaur, what are the features that confirm this identification? A classification is certain if it is based on characters considered to be diagnostic—that is, typical and restricted to a group of organisms. So-called primitive characters, understood as basic features (e.g., eyes, spinal column), have poor diagnostic value, because they are very common; "derived" characters, which evolved secondarily (e.g., birds' feathers, horses' hooves), are more useful because, since they are rarer, they narrow the possibilities of choice and direct the researcher toward a more precise relationship. To understand the links between living beings, it is necessary to seek those individuals that share derived characters because they have been handed down from a common ancestor.

*Scipionyx* reveals numerous derived characters in common with the suborder of theropod dinosaurs. Like them, it has sharp, pointed, jagged-edged teeth; a special joint between the jawbones; a narrow, elongated shoulder blade; two wrist bones (carpals) fused into a single element that "embraces" the ends of two bones of the hand (metacarpals); forelimbs with three fingers armed with claws; and pubic bones ending in a foot-shaped expansion. Among the theropods we can then recognize *Scipionyx* as belonging to the infraorder of the coelurosaurs because it has numerous very particular derived characters in common with them, including forelimbs that measure three-quarters of the length of the body, an ulna (a bone of the forearm) that forms a backward-bent arch, and a first meta-

Two young *Scipionyx* compete for a *Chometokadmon,* a small lizard-like reptile whose fossil remains were found in the fossil site of Pietraroia and which was probably one of the favorite prey of these agile coelurosaurs. *Drawing by F. Fogliazza.*

carpal bone that measures one-third the length of the second. Many coelurosaurs can in turn be grouped into a subgroup called maniraptoriforms. Among the characteristics they have in common, some (such as elongated anterior articulations of the cervical vertebrae and a crescent moon–shaped carpal bone) can also be found in *Scipionyx.*

These are detailed anatomical characteristics, many of which are not even visible to the naked eye because the dinosaur of Pietraroia is really tiny. Its skeleton measures a mere 23.7 centimeters, and its head just a little more than 5. Taking into account the missing parts, the complete animal was probably not more than 50 centimeters long, and in life weighed about half a kilo. Dinosaurs this small that are not embryos still in the shell can be counted on the fingers of one hand. Many people will wonder how we can be so certain that *Scipionyx* is a hatchling. Here too, the proof is found in a number of anatomical details.

First and foremost, the skull of the dinosaur of Pietraroia is larger, compared to the body, than in any other adult theropod, including the *Tyrannosaurus rex.* "Ciro" had a really big head, a short snout, and big, round eyes, which made him in some sense "cute." Paleontologists have observed that these proportions are found in a number of rare dinosaur embryos discovered inside particularly well-preserved eggs. Study of *Scipionyx* confirms that newly hatched dinosaurs, unlike the young of fishes, amphibians, and some reptiles, did not resemble their parents but possessed those infantile anatomical characteristics that in adult birds and mammals induce aggression-blocking behavioral reactions and care and protection of their offspring. In other words, they were real "babies."

Very little is yet known about the growth (ontogeny) of carnivorous dinosaurs, because rarely do we have a sufficient number of fossils to be able to recognize the stages of growth. Exceptions include the little coelophysoids *Coelophysis* and *Syntarsus,* which lived in very large herds. "Cemeteries" of these theropods have been found in New Mexico and in South Africa, containing hundreds of skeletons of young and adult individuals, which provide evidence of differential growth of the various parts of the skeleton (allometry): the body proportions changed with age. From comparison with these data, it can be inferred that as an adult, *Scipionyx* would have had a longer snout and anterior portion of the jawbone, proportionally smaller eyes, a longer neck, and shorter terminal segments of the limbs, ending up closely resembling a *Velociraptor.*

But there are even more interesting details which prove that "Ciro" was just a little larger than a hatchling.

If *Scipionyx* had been a mammal, it could be said that it still had its milk teeth. Like all reptiles, dinosaurs replaced teeth continuously; replacement did not occur at random but in successive waves, according to a precise order. In the little Italian dinosaur, similarity in the size of the teeth placed in corresponding positions on the left and right maxillary toothrows suggests that the first wave of replacement had not yet occurred. In dinosaur embryos about to hatch, some teeth erupted sooner than others, and they evidently did so in the same positions on both sides of the jaws. Then the teeth that emerged last reached the size of the first ones, and within the space of a few weeks the tooth set in each jaw was composed of a row of identical points. Since teeth were replaced more or less at random, the mirror symmetry between the tooth sets in each jaw would have been lost within a few months after birth. The most likely explanation for *Scipionyx*'s having its original set of teeth is that it was recently hatched.

The little denticles that give the cutting edges of theropods' teeth their typical jagged appearance may also help to reveal the animal's stage of development. "Ciro" has very few denticles compared to the teeth of adult carnivorous dinosaurs, while embryonic theropods have very few or no denticles at all.

Several bones in *Scipionyx*'s postcranial skeleton are not yet fused: the scapula and coracoid (bones of the pectoral girdle) remain distinct, and several vertebral bodies and their respective neural arches (spine-like processes of the vertebrae) are separated or even disarticulated. The sternum and sternal plates, which were not fossilized, must still have been cartilaginous. All of these features suggest a very young age.[12]

So was the baby *Scipionyx* weak, defenseless, and unable to walk? Did it have to stay in the nest, depending upon a parent to provide it with food and protection? No, the little dinosaur was perfectly self-sufficient. "Ciro's" case is similar to that of "precocious" chicks: the limbs become ossified before the eggs hatch, while the vertebrae fuse much later. But that does not prevent the hatchlings from supporting the weight of their body; even with an incompletely ossified spinal column, they are perfectly able to walk about.

It is interesting to note that in birds, the bones of the wrist and the sternum ossify at a later stage—the same stage at which the vertebrae fuse. By comparison, *Scipionyx*'s well-formed carpus is decidedly precocious. This means that it could move its wrist joints perfectly right from birth, which gives us reason to believe that little *Scipionyx* needed to do so in order to capture food.[13]

"Ciro's" very tender age is one of the reasons why it is not possible to know whether it was male or female. In reptiles and birds, sexual characteristics are not evident at this stage of development. Then again, there is no other *Scipionyx* with which to make comparisons. The sex of dinosaurs is one of their lesser-known aspects. In some adult dinosaurs, such as the duck-billed dinosaurs, there are evident differences in the size of the crests adorning the skull, which have been construed as secondary sexual characteristics. In the tyrannosaurs and other carnivorous dinosaurs, it would appear that the females had fewer spine-like processes (chevrons) under the first vertebrae of the tail than the males (so as not to obstruct the passage of the eggs).

Because very little is known about the growth patterns of the dinosaurs, it is difficult to say exactly how big *Scipionyx* would have grown and how long it would have taken it to reach adulthood. Present-day reptiles grow at an indeterminate but discontinuous rate; that is, they increase in size throughout their life, growing faster during the warm

A close-up of the skull of *Scipionyx*. The large circular eye sockets, short snout, and symmetrical length of the teeth along the right and left maxillary arches indicate that the animal was still very young when it died. *© Italian Ministry for Cultural Heritage and Activities, Archeological Superintendence of Salerno, Avellino, and Benevento. Photo by L. Vitola. Reprinted by permission from* Nature *392, no. 6674. © 1998 Macmillan Publishers Ltd.*

season and slower during the cold season, because, being "cold-blooded" (ectothermic) animals, their metabolism accelerates or slows down in accordance with environmental temperature. Food availability and quality also have a big impact on growth rate. Upon sectioning the bones of some species of dinosaurs and observing them under the microscope, paleontologists have found growth rings, similar to those of present-day reptiles, which would indicate seasonal growth rhythms. On the other hand, recent histological examinations conducted on the bones of young theropods of the genus *Syntarsus* show that their bone tissue was more highly vascularized than in the adults, which suggests that their growth must have been rapid in the first few years of life, as in birds and mammals, and then slowed considerably. This incongruence leads us to presume that dinosaurs followed a peculiar physiological pattern, perhaps without present-day equivalents. If it were possible to take a look at a bone sample of the dinosaur of Pietraroia, we would have a clearer idea

A life-size reconstruction of the baby dinosaur, which was probably no more than 50 centimeters in length and weighed half a kilo. © *Natural History Museum of Milan. Model by F. Fogliazza. Photo by L. Spezia.*

of how much it could have grown. However, considering that there is a high degree of ossification of "Ciro's" numerous skeletal parts, we can estimate that an adult *Scipionyx* was not more than 1.5 meters in length and weighed around 25 to 30 kilos. Having a "baby" carnivorous dinosaur at our disposal is a true rarity; precisely for this reason, even though samples of bone tissue can provide a lot of new information, the expediency of carrying out "destructive" tests on the little Italian dinosaur should be considered with the utmost caution.

## An Extraordinary Mosaic

Besides its tender age, a second feature that caught the interest of paleontologists around the world is "Ciro's" mosaic of osteological characteristics. In comparison to the already known carnivorous dinosaurs, there is very little that is absolutely new: a pair of particularly well-developed maxillary teeth, which make it look somewhat like a vampire; a dentary bone that is longer than the maxillary bone; and a wrist composed of only two compressed lens-shaped bones positioned one on top

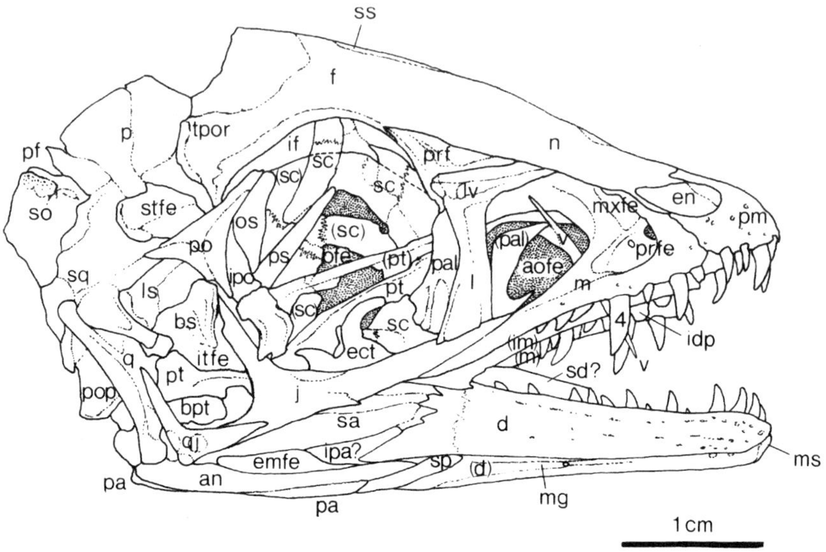


A sketch of the skull of *Scipionyx*. Abbreviations: an—angular; aofe—antorbital fenestra; bpt—basipterygoid; bs—basisphenoid; ect—ectopterygoid; emfe—external mandibular fenestra; en—external naris; f—frontal; idp—interdental plates; if—inner (orbital) wall of frontal; im—inner (lingual) wall of maxillary; ipa—inner prearticular; ipo—inner (orbital) wall of postorbital; itfe—infratemporal fenestra; j—jugal; l—lachrymal; ls—laterosphenoid; lv—lachrymal vacuity; m—maxillary; mg—meckelian groove; ms—mandibular symphysis; mxf—maxillary foramen; n—nasal; ofe—orbital fenestra; os—orbitosphenoid; p—parietal; pa—prearticular; pal—palatine; pf—parietal flange; pm—premaxillary; pno—pneumatic opening; po—postorbital; pop—paroccipital process; prf—prefrontal; ps—parasphenoid; pt—pterygoid; q—quadrate; qj—quadratojugal; sa—surangular; sc—sclerotic plates; sd—supradentary; so—supraoccipital; sp—splenial; sq—squamosal; ss—sagittal suture; stfe—supratemporal fenestra; tpor—transverse postorbital ridge; v—vomer; 4—fourth maxillary tooth. Left-side elements are in brackets. Scale bar 1 cm. *From Dal Sasso and Signore 1998. Drawing by F. Fogliazza. Reprinted by permission from* Nature *392, no. 6674. © 1998 Macmillan Publishers Ltd.*

of the other. What really makes *Scipionyx* one of a kind is the combination of characters—attributes that had previously been observed separately in different families of coelurosaurs, and which, precisely for this reason, constitute their distinguishing features.

The dinosaur of Pietraroia initially appeared to have a larger number of original characters. It was hypothesized that the dinosaur had a crest on its forehead, but then it was realized that the left and right frontal and nasal bones had been flattened against each other during compaction of the sediment and subsequent fossilization. Another somewhat extravagant hypothesis was that the two larger maxillary teeth, which seemed to have two fine symmetrical channels running through them, were poison fangs; the channels later turned out to be fractures that by pure chance happened to be identical.[14]

But let us take a look at the cocktail of characters that make *Scipionyx* a truly unique dinosaur.

The skull and lower jaw bones are in part organized as in the dromaeosaurids *Deinonychus, Dromaeosaurus,* and *Velociraptor.* The shape and proportions of the forelimbs are also very similar to those of the agile and voracious predators made famous by *Jurassic Park.* Despite its immaturity, the forelimbs are highly developed in the theropod of Pietraroia, probably because they were essential for predation right from the first days of life. Other cranial characters, plus the configuration of the cervical vertebrae and the orientation of the pubic bones, resemble those of the troodontids, a family of highly evolved coelurosaurs similar in appearance to the dromaeosaurids but perhaps more closely related to the bizarre *Oviraptor* and the huge *Tyrannosaurus.*

By contrast, the bones of the pelvic girdle in *Scipionyx* have numerous characters that are found in the so-called ostrich-dinosaurs (the *Gallimimus* of Spielberg's film), which make up the family of the ornithomimids.

At the same time, it is important to note that the little Italian dinosaur differs from the maniraptoriforms ("dinosaurs with predatory hands," the group to which the above-mentioned families belong) because it retains many unequivocally primitive features typical of basal coelurosaurs such as *Compsognathus, Sinosauropteryx,* and *Ornitholestes.* These characters are a well-developed prefrontal bone, a stout lacrimal bone, an anterior protrusion of the scapula (the acromion process), a coracoid that is rounded and not square, the presence of a terminal expansion of the ischium, the absence of a muscle attachment on the femur (referred to as posterior trochanter), and the fine V-shaped bones located under the caudal vertebrae, referred to as chevrons, which are slender and lack expansions.

The latter two features, which are seemingly of little importance, are, on the contrary, significant clues that led us to exclude the initial hypothesis that *Scipionyx* possessed the typical adaptations of the "raptors." Dromaeosaurids and troodontids, which are the most highly evolved maniraptoriforms, were characterized by a functionally didactyl foot (that is to say, they walked on only two toes) because they had a specialized internal toe, which was kept raised from the ground, for predation. In the dromaeosaurids, in particular, this toe was constructed in such a way as to ensure the utmost mobility of the famous scythe-like talon with which it was armed. The talon was connected to a phalanx by a very wide articulation, which enabled it to rotate downward by more than 180 degrees and impale the flesh of its prey with a rapid movement. Although the tail, which is reinforced on the central and terminal portion (but not at the base) by rod-like extensions of the vertebrae and the chevrons, was less flexible than in other theropods, it behaved elastically and could be used for balance when running and jumping. According to some paleontologists, it could even be used as a prop: the attack was made by kicking, and in this case the animal could lift both hind legs and drive the two talons into the body of its prey while supporting itself on its tail. *Velociraptor* and the like consequently had a special muscle used for withdrawing the hind leg without involving the caudifemoral muscle (the latter, being responsible for supporting the tail, would otherwise have transmitted troublesome oscillations at each "kick," causing a loss of balance at the crucial moment of the attack). That special muscle, called M. ischiotrochantericus, leaves a definite trace of its insertion (posterior trochanter) on the rear edge of the femur, which in *Scipionyx* does not exist. While the "normal"-shaped chevrons, being too close to the base of the tail, did not demonstrate anything, the absence of the posterior trochanter proves—despite the fact that a large part of the hind limbs is missing—that the foot of the dinosaur of Pietraroia must have been of the generic type, with three identical toes that rested on the ground, and thus was more primitive than that of the "raptors."

Other important information that enables us to understand "Ciro's" functional anatomy, and consequently its lifestyle, can be obtained from the structure of the bones of the wrist and hand. Although the general proportions of the forelimbs are very similar to those of the dromaeosaurids, the extremities are a little different. The metacarpal bones are as long as the forearms. This is very long in comparison to all other coelurosaurs and is thought to have something to do with the young age of the animal. In addition, the first digit of the hand is much shorter than the second and third compared to its length in the maniraptoriforms, and in this it is more similar to that of compsognathids. In any case, the first digit is not opposable as our thumb is, because at the base it is closely coupled with the second digit, as is the latter with the third. The "palm" of the hand was consequently rigid, and this prevented *Scipionyx*'s limbs from having prehensile capabilities.[15] The articulations of the wrist and

Detail of *Scipionyx*'s right hand. Note the presence of the horny sheaths that still cover the terminal portion of the bony talons. As a rule, these structures, which are composed of keratin and are thus more fragile, do not become fossilized. *© Italian Ministry for Cultural Heritage and Activities, Archeological Superintendence of Salerno, Avellino, and Benevento. Photo by L. Vitola.*

phalanges therefore served to clutch prey not so much with one hand but rather by using one hand against the other, a bit like squirrels do with peanuts.

The most important conclusion that emerges from this detailed examination of data is that because of its mosaic of anatomical characters influenced by juvenile features, *Scipionyx* cannot be easily classified. There is no doubt that the little theropod of Pietraroia bears witness to the existence of a new kind of coelurosaur. However, a new family cannot be created unless it is found that a new genus cannot be included within the range of variability of an already existing family (in the case of *Scipionyx*, the most likely family is the Compsognathidae). Moreover, the International Code of Zoological Nomenclature states that in order to establish a brand new family, it is first necessary to find another specimen belonging to a second closely related genus.

But there is another implication that should not be overlooked: in evolutionary terms, how can an animal incorporate anatomical characters present in different families that lived in different periods and places? The most logical and simple explanation for this cocktail of characters would be that *Scipionyx* represents the common ancestor, or rather a species forming part of the stock from which the aforementioned coelurosaurs subsequently evolved—a sort of "prototype." In actual fact, the matter is a little more complicated, because the fossil record tells us that 110 million years ago, all the families of coelurosaurs were already differentiated—and we are absolutely certain about the dating of the Pietraroia fossils because the extremely fine marine limestone of the Benevento deposit contains microfossils typical of the Albian, the uppermost Lower Cretaceous.

So how, then, could a theropod of the Cretaceous Period have preserved primitive characteristics already present 50 million years previously in Jurassic coelurosaurs and at the same time have evolved in other anatomical parts according to the "fashion" of its own era?

## Islands and Genes Adrift

It looks as though "Ciro" followed his own path—that he evolved in a similar way to the little Cretaceous theropods of the rest of the world but, with Latin originality, going a little adrift.

Prolonged geographical isolation could have favored the preservation and expression of ancestral genes in *Scipionyx*'s forebears. At the same

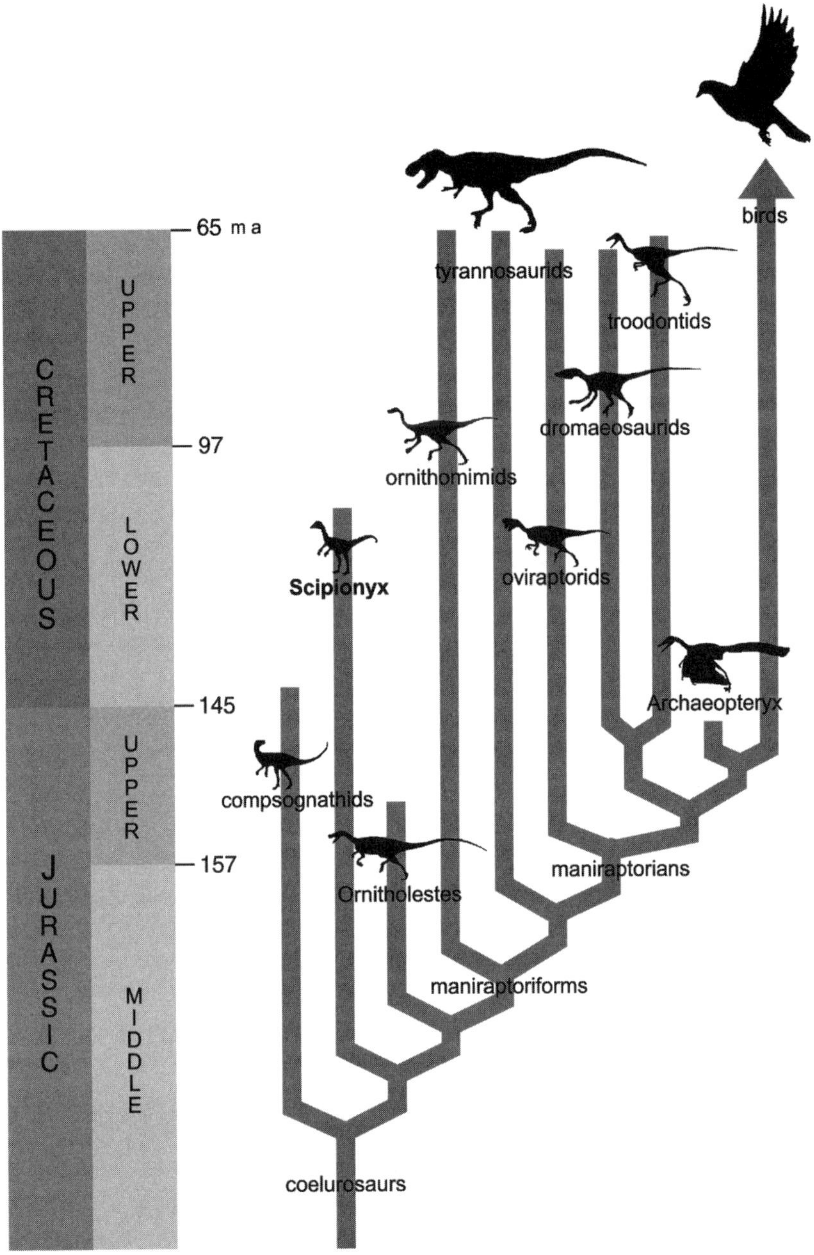


Hypotheses on the relationships between *Scipionyx* and the other coelurosaurs, displayed in a simplified cladogram based on the most recent theories. The dinosaur of Pietraroia may represent a separate lineage of carnivorous dinosaurs. *Graphics by F. Nodo.*

time, this archaic set of genes would have possessed the same capacities for development of those anatomical characteristics that on the continents led to the differentiation of the families of ornithomimids, troodontids, and dromaeosaurids. Starting from this DNA, isolation would have favored variations on the theme of those same characteristics plus the appearance of original elements. By adaptive convergence, natural selection would have then favored the acquisition and preservation of some of the modifications also seen in the other coelurosaurs of the Cretaceous period. However, for this to be more than a fantastic hypothesis, there has to be paleogeographic confirmation that *Scipionyx* did evolve in isolation.

Through the research of geologists, we know that during the first half of the Cretaceous, the climate was hot and humid all over the earth. The poles were devoid of ice caps, with temperatures ranging from 0 to 15 degrees centigrade, and the seasons were not as clearly defined as today. This seems to be related to the fact that the Cretaceous was a time of intense tectonic and volcanic activity. High concentrations of carbon in the sediments of this period prove that volcanism released huge quantities of carbon dioxide into the atmosphere, causing a greenhouse effect similar to the one currently produced by human activities through the burning of fossil fuel. The intense seismic and volcanic activity was a sign that the supercontinents were beginning to break up. New fractures opened up between the plates of continental crust, giving origin to new ridges, which belched out enormous quantities of magma and gas. Flooding these rifts, new seas worked their way between the stretches of dry land, oceans in embryo that gradually widened, separating Africa from South America, India, Antarctica, and Australia.

This change in the distribution of land and sea and the consequent isolation of many regions of the earth accelerated the diversification of the living creatures, including the dinosaurs. Geographic isolation is the main cause of speciation because populations that had once exchanged genes were literally cut off from one another. As a result, each of them began to evolve along its own path. The most striking example is Australia, the huge island continent. The fact that it broke away from the other lands before the establishment there of placental mammals enabled marsupials to survive and evolve independently.

Isolation is the most probable geographical condition in which paleo-Italy found itself during the Lower Cretaceous Period in the middle of the Tethys, the long sea that crossed half the world in an east-west direction, slightly above the Tropic of Cancer, separating Eurasia (to the north) from Africa. Because of the presence of this sea, until a few years ago it was thought that Cretaceous Italy was completely submerged and had consequently never been dinosaur land. Evidently, however, the Apennine Platform, corresponding to part of southern Italy, must have been above sea level and large enough, at least in some areas, to provide habitat for dinosaurs.

It should be remembered that we are talking about a carbonate platform, a geological structure devoid or almost devoid of relief, a plateau of continental crust covered by a sea, generally not very deep but extensive, which geologists describe as "epicontinental." The shallow depth of epicontinental seas enables marine organisms that need sunlight for their vital processes (especially corals, sponges, and calcareous algae) to fix a large amount of calcium carbonate in their skeletons. After the organisms die, the calcium carbonate accumulates in large quantities on the seabed due to disintegration of their skeletons and chemical precipitation—hence the term "carbonate platform." So "platform" does not mean dry land, but the water is usually so shallow as to allow islands and peninsulas to emerge here and there. In the present-day Bahamas, for example, plants and terrestrial animals exist because of a difference in level of a few meters at most.

Carbonate platforms are highly sensitive to changes in sea level. It takes only a meter of water, more or less, to submerge or expose dozens of square kilometers of coastline. Vast regressive episodes—drops in sea level with consequent emergence of strips of land—are documented for the time in which "Ciro" lived (the Lower Albian stage, corresponding to approximately 110 million years ago) on the Friuli Platform and in the Matese area.[16] The area of Pietraroia consequently formed part of the Apennine Platform, a small fragment of Africa elongated from east to west and considered to be a separate paleogeographic unit because it was isolated from the adjacent Apulian and Latian platforms by narrow but very deep basins.[17] Compared to the nearby Apulian and Dinaric platforms, the Apennine Platform was of very limited dimensions, no wider than about 100 kilometers and no longer than 500. Bearing in mind that only part of a platform can be above sea level—and in our case that would mean having an area smaller than that of Corsica—we can imagine two probable biological consequences: in an isolated limited area, it is highly likely that there would have been endemic species (exclusive to that geographical area) and that there may have been other effects of insularity such as dwarfism. The new kind of theropod represented by

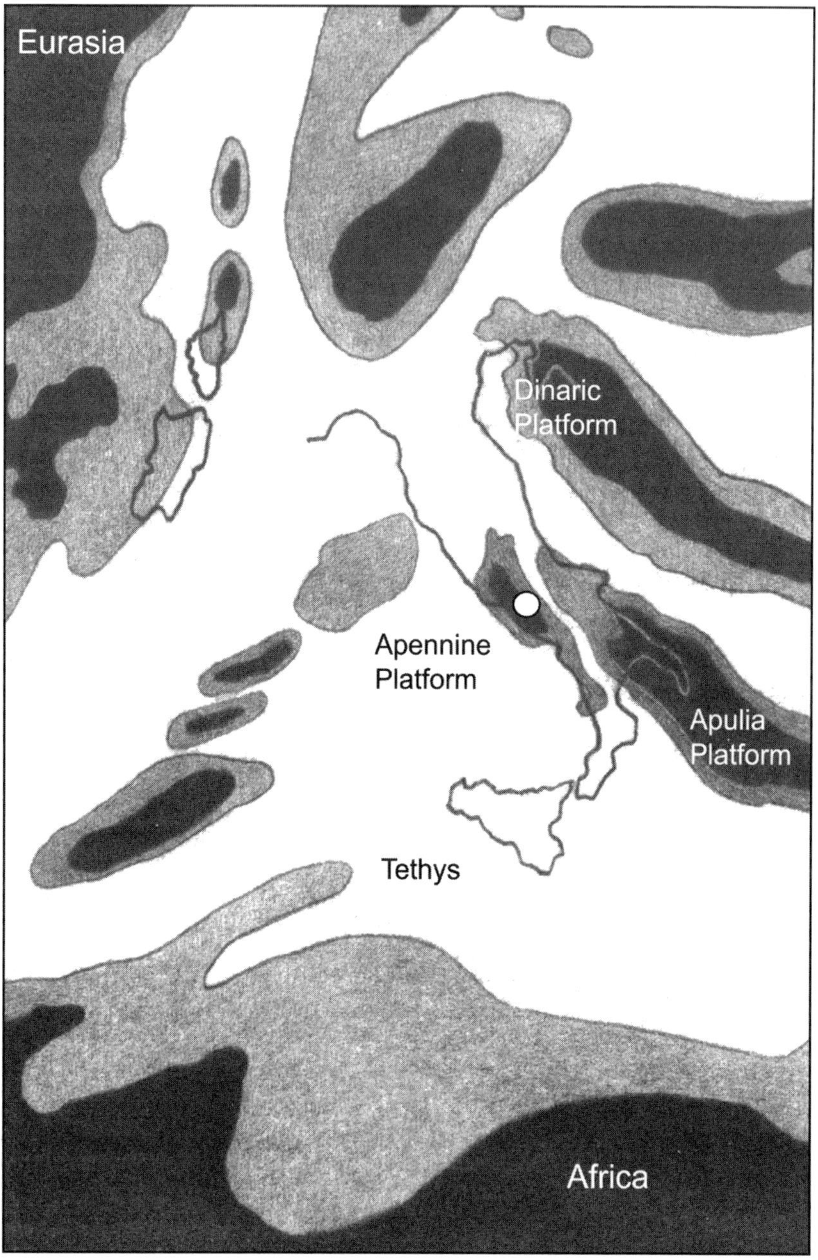


Italy at the time of *Scipionyx*. A large part of the Italian peninsula was submerged by the Tethys Sea (white) and was broken up into various carbonate platforms (gray). These were partially above sea level (black) and permitted the existence of terrestrial plants and animals, including dinosaurs. The area where Pietraroia is now located was on the Apennine Platform, indicated here by the white dot. *From Zappaterra 1990, modified. Drawing by F. Fogliazza.*

*Scipionyx* could quite easily be an endemic form. So it would come as no surprise if in the future no other *Scipionyx* or any other closely related dinosaur is discovered anywhere else in the world. As we already knew, dinosaurs were not all huge. Perhaps the Apennine Platform during certain periods was in turn divided into an archipelago of islands with even smaller strips of land, and perhaps *Scipionyx* was a species of dwarf dinosaur. Dwarfism also appears to characterize the crocodiles discovered at Pietraroia, which measure no more than a meter in length.

Evolution, which often retraces its steps, produced dwarf elephants in the Mediterranean area and on islands even larger than Corsica in recent geological times. *Elephas falconeri,* which lived in Sicily, Malta, Crete, and Cyprus in the Pleistocene, evolved as a result of geographical isolation of various populations of *Elephas antiquus,* a pachyderm that on the continent was at least 4 meters tall. The dwarf species, in contrast, was no more than 1 meter in height. Another example of insular dwarfism is the pony, a tiny horse now bred all over the world but native to the Shetland Isles.

Dwarfism is evolution's response to lack of space and food. Under prolonged conditions of isolation, in limited distribution areas, a species can survive by only one strategy: becoming smaller. It is better to be numerous and small rather than few and large. In fact, a larger number of individuals have more probability of surviving, not only for statistical reasons but also for a genetic reason—that only in this way is it possible to maintain the genetic variability needed to ensure adaptation to new environmental conditions. The viability of a species is measured by its genetic variability.

## Internal Organs

Leaving aside these conjectural, albeit well-grounded, considerations, we now come to the icing on the cake: *Scipionyx* retains soft tissues and anatomical details never seen before in any other dinosaur in the world, first and foremost including several internal organs perfectly fossilized in three dimensions.[18]

Integument traces are known for several kinds of dinosaurs. Skin impressions have been found on several skeletons of North American hadrosaurs. Several fossilized muscle fibers were attached to the bones of a Brazilian theropod, and the remains of integumentary structures were observed in the Argentine abelisaurid *Carnotaurus,* the Spanish ornitho-

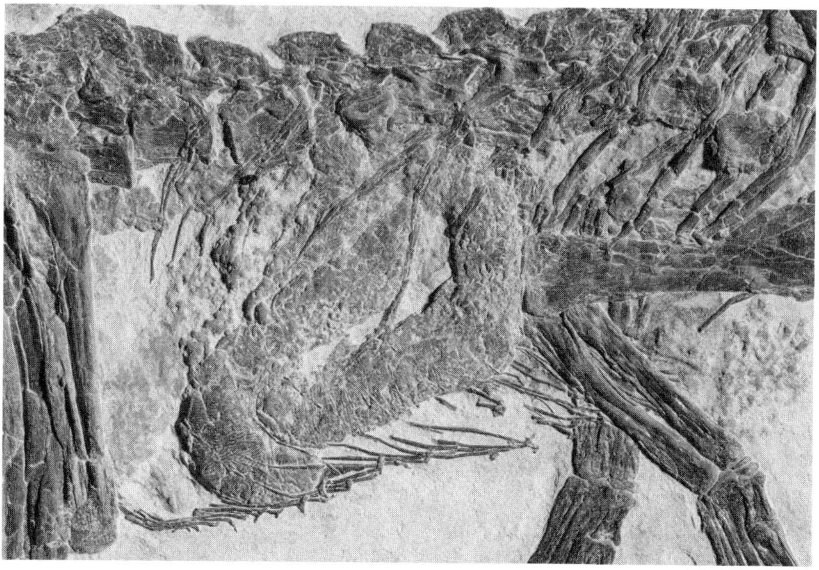

In *Scipionyx* the intestine has been preserved—the only case of its kind in the world. Its fossilization is perfect, to the point that the folds of the mucosa layer are also visible. The reddish patch located between the forelimbs (to the right in the photo) would appear to be the remains of the liver. *Reprinted by permission from* Nature *392, no. 6674. © 1998 Macmillan Publishers Ltd.*

mimid *Pelecanimimus,* and the "feathered" dinosaurs discovered in China beginning in 1995. In April 2000, the discovery of a fossil heart was announced; the putative heart was preserved as a nodule of iron minerals inside the thorax of a herbivorous dinosaur (belonging to the genus *Thescelosaurus* but known by the nickname of "Willo") extracted from the Cretaceous deposits of South Dakota.[19]

None of these specimens, however, provides as much information about the soft anatomy as "Ciro."

The organ that appears to be the most complete in *Scipionyx* is the intestine, which, to use the words of the English paleontologist Mike Benton, is clearly visible, "lumpy and shiny, almost as you would see it after dissecting a modern animal."[20] Its fossilization is perfect at the microstructural level as well, to the point that in the loops of the digestive tract you can see the folds of the mucosa (the innermost muscular layer), which measure a few tenths of a millimeter but are clearly visible even to the naked eye, with their crosswise pattern and with typical interlac-

Tendons (*left*), bundles of muscle fibers (*center*), the terminal portion of the intestine (*right*), and the remains of the muscles of the ischium (*right margin of the photo*) are preserved at the base of the tail. © *Italian Ministry for Cultural Heritage and Activities, Archeological Superintendence of Salerno, Avellino, and Benevento. Photo by L. Vitola.*

ings, similar to the ripples of sand on the seashore, called anastomoses. The organ has not yet been studied in detail. Since it folds over upon itself several times, it is not easy to follow its precise course; however, the surface of the digestive tract appears to be more irregular in the first section and smoother in the rear section, and it seems reasonable to presume that the unevenness is due to the presence of alimentary boluses (swallowed food) not yet liquefied. A large part of the intestinal tract consists of the small intestine, which measures on average 5.22 millimeters in diameter. A very important feature is its location; the intestine is positioned much farther forward in the abdominal cavity than was thought, and consequently it is not supported by the pubic bones.

Since soft tissues by their very nature are quite mobile, it is always difficult to be certain that they retain their original position in fossils. However, "Ciro" was not run over by a truck but was very slowly and plastically deformed by the pressure of the sediment. Besides that, proof that the intestine is *in vivo* position is provided by the so-called gastralia.

In the dinosaur of Pietraroia, these fine, fragile abdominal ossifications, which are usually dispersed in the sediment, are all perfectly aligned. What is more, several gastralia closely embrace a ventral loop of the intestine, unequivocally revealing, for the first time ever, their contribution to the effective support of the gut.

At present it is not clear what was in the space in front of the pubic bones. According to some experts, such as the American Gregory Paul, besides the urogenital organs this was the site of abdominal air sacs, "balloons" essential for ventilating the lungs of the respiratory system of birds and perhaps already evolved in certain theropods.[21]

While the small intestine is tangled around itself, *Scipionyx*'s colon runs dorsally beneath the vertebrae in the lumbar region, parallel to them (a sign that it was "strung up" to the roof of the abdominal cavity by means of a special connective tissue called mesentery), and passes through the pelvic cavity between the two right and left arms of the V-shaped pubic bones. Last, it runs down in spirals, which rest on the bones of the ischium, and ends immediately above their extremities. The dinosaur's cloaca (the common aperture of the digestive, urinary, and reproductive systems) is also visible.

Absorption of nutritive substances is one of the most important metabolic functions of an organism. Since shape indicates function in biology, and since one function is always linked to another, by studying a fossil organ it may be possible to deduce its function. Anatomical study of the digestive tract of a dinosaur could therefore be of substantial help in learning about its physiology. At the moment it is too soon to promise a result, but perhaps "Ciro's" fossil intestine will help us, as "Willo's" heart did, to address with anatomical evidence the long-disputed question of whether dinosaurs had warm blood or cold blood. What we can say so far is that *Scipionyx*'s digestive tract is surprisingly short and stumpy, a sure sign of an animal protein–based diet.[22]

Besides the intestine, *Scipionyx* shows exceptional three-dimensional fossilization of bundles of striated muscles that run from the base of the tail. These bundles, which extend over an area of a couple of square centimeters, are perhaps the remains of one of the two large muscles that supported the tail along the sides, the caudifemoral muscles. Another possibility is that they belong to the cloacal muscle, which is an unpaired muscle located ventrally along the plane of symmetry of the body. Whatever their precise identity may be, these muscle fibers could prove to be of the utmost interest because the fossilization is so perfect as to allow us to distinguish the individual cells. By studying their microstructure, it may be possible to acquire information about their motor capacities and, hence, obtain other invaluable data on the physiology of dinosaurs.

Another area in which muscle remains can be observed is the dinosaur's breast. Here, however, the individual fibers are not united in bundles but are broken up into a haphazard array of microscopic needle-like cells that are visible only at magnification of more than 50X. For this reason, the pectoral muscles were initially thought to be part of the dinosaur's skin.

At the center of this triangular-shaped area there are seven or eight light-colored, almost diaphanous rings partly immersed in the pectoral muscles. It caused a great stir when, during the initial phases of the study on the fossil, it was discovered that they were exactly where the trachea (the respiratory tube) in dinosaurs should have passed, namely, in the prethoracic region, in line with the plane of symmetry of the bones of the pectoral girdle and slightly above the furcula.[23] In fact, these rings, which are placed one on top of the other and are cartilaginous in appearance and U-shaped, could be nothing other than part of the tracheal tube. In reality, the rings do not in fact appear to be fully closed but seem to be incomplete, with the dorsal arch not chondrified (i.e., consisting of cartilage that is not yet well-formed). It is not by chance that this morphology is found in the tracheal rings of young individuals of many terrestrial vertebrates.

It is highly likely that the reddish patch located between *Scipionyx*'s forelimbs is the remains of the liver. Apart from the anatomical position (the liver generally lies beneath the heart, which in the dinosaurs, as in many vertebrates, had to be protected by the sternum), the main clue to this interpretation was the color. That coloring is typical of hematite, a mineral whose formation requires a lot of iron, and which, as its name curiously appears to suggest, in the Italian dinosaur could have derived from an accumulation of blood.[24] And there is more blood in the liver of an animal than in any other organ.

Last, another extraordinary fact is that *Scipionyx* has retained its actual claws—the horny portion of the talons, composed of keratin, which still covers the nail phalanges. These delicate structures have been preserved because they were covered with an extremely fine layer of matrix. Although keratin is a highly resistant insoluble protein, there are only a handful of fossil specimens in which it has been preserved. The better-known ones are *Archaeopteryx* of the Bavarian Jurassic and *Pelecanimimus* from Spain. An expedition by the American Museum of Natural History of New York recently found horny sheaths also on the nail phalanges of several *Oviraptor* skeletons extracted from the Cretaceous deposits in Mongolia. This discovery proved to be even more important when, upon analysis of the antibodies, it was discovered that the molecular structure of the keratin preserved in these specimens is identical to that of birds' feathers and not to that of reptiles' scales. But once again *Scipionyx* beats them all: here the preservation is so perfect that the fine layers of keratin covering the back of the nails are darker in color than those covering the ventral portion. The deep color could mean higher keratin density, or strength, and this would also explain why the tips of the talons are almost black.

Keratin is the first original organic substance we note as being present in *Scipionyx*. Until now we have spoken about minerals—inorganic substances that, as a rule, completely replace organic remains during the fossilization process.[25] Beginning in the early 1990s, working with some exceptionally well-preserved fossils and using increasingly sophisticated analytical equipment, paleontologists were able to detect the presence of different types of organic molecules, especially residues of proteins and lipids. The American paleontologist Mary Schweitzer, now at North Carolina State University but previously a close colleague of Jack Horner at the Museum of the Rockies in Bozeman, Montana, is a specialist in this type of research. She has demonstrated that although the almost perfect preservation of form does not guarantee the presence of original organic material, the preservation of endogenous molecules is closely correlated to it. The fossilization of different types of tissues in different colors also suggests different chemical composition. Schweitzer therefore believes that in *Scipionyx* there are organic molecules that belonged to the cells of the dinosaur: for example, there might be hemoglobin molecules or their decomposition products in the reddish pigments of the liver, while in the muscle fibers it may still be possible to find myoglobin and actin.

According to David Vesely, an endocrinologist at the University of Florida, *Scipionyx*'s intestine could contain atrio-natriuretic peptides, vasodilator hormones that are usually released by the heart but are also well concentrated in the gastrointestinal tissues.

A moot point is the presence of nucleic acids (DNA, RNA), found with certainty in fossils up to 15–20 million years old. For several years, a dozen laboratories scattered in various countries have been attempting to extract DNA from dinosaur bones, teeth, and eggs. On various occasions, successful outcomes have been announced, but in none of the cases has it been possible to repeat the operation on the same specimen in another laboratory. However that may be, even if one day we were to find traces or fragments of dinosaur DNA, we can forget the idea of being able to use it to bring dinosaurs back to life. Unfortunately—or rather, luckily—cloning a dinosaur is and will always be impossible.

In April 2000, an American private research company announced to the world that it had managed to decipher, for the first time in history, the entire sequence of the human genome. The nucleotides, the molecules that make up the structural units of the DNA of *Homo sapiens,* correspond to approximately 3 billion letters, which, when written one next to the other, would fill eighty volumes of a thousand pages each. Since it must contain equally complex information, the DNA of other vertebrates, including dinosaurs, cannot be much shorter than human DNA. Even hypothesizing that the processes of organic degradation have permitted the preservation of a few original dinosaur cells and their nuclei containing DNA, the thermal and pressure shocks that they have been subjected to for millions of years have undoubtedly caused the decomposition (denaturation) of the proteins that hold together the double helix of the DNA and, consequently, its disintegration and fragmentation. Assuming that even after all these transformations we were able to isolate uncontaminated fragments of original dinosaur DNA, even in the best of cases the percentage of nucleotides in correct sequence would be so low as to be equivalent to a few hundred words in all, scattered at random through the eighty thousand pages of the volumes. How can we expect to know the contents of the entire encyclopedia from these few scattered words? Extinct species, or at least "Ciro" and his friends, are gone forever.

Besides molecular studies, *Scipionyx* could also be subjected to histological examinations, the results of which could generate a great deal of discussion and involve numerous disciplines such as physiology, comparative anatomy, immunology, and molecular biology. Such work is only the beginning. As Lawrence Witmer, paleontologist and anatomist at Ohio University, wrote to the author of this book, "This is a specimen that the community will have to grow with."

After this journey into the very small, where inside a fossil a mere 23 centimeters long we discover a whole universe of molecules and the perfection of anatomical details of what was once a living creature, we naturally ask how such a specimen could have been preserved so well.

The quick answer is: precisely because it was tiny, and therefore it was quickly and easily buried by the sediment that prevented it from destruction. However, the real secret lies in a combination of very unusual environmental conditions that make Pietraroia a truly extraordinary location.

## The Secret Is in the Lagoon

The paleontological outcrop of Pietraroia is considered to be a *Konservat-Lagerstätte,* a "site of exceptional fossil preservation." As a rule, only the hard parts of organisms—the bones, teeth, or shells—become fossilized. In *Lagerstätten,* which are extremely rare, very particular environmental conditions can also allow the preservation of soft parts—the skin, muscles, and internal organs of a vertebrate—or even the entire body of an invertebrate lacking an internal skeleton, such as a jellyfish or a worm. More precisely, the term *Lagerstätten,* coined by the German paleontologist Adolf Seilacher, indicates "sedimentary bodies which provide an unusual quantity of paleontological information" that were

The fossil site of Pietraroia is composed of stratifications of extremely fine-grained yellowish-gray limestone formed on the bottom of a tropical sea with calm, shallow waters in an environment very similar to the present-day Bahamas. *Photo by C. Dal Sasso.*

An inland lagoon on Bimini Island (Bahamas). *Photo by C. Dal Sasso.*

formed in closed, stagnant basins in which deposition involved for the most part very fine-grained sediment.

This is a very theoretical definition. An exceptional type of fossilization can be facilitated by numerous factors, and in reality, things do not occur according to rigid patterns. Each fossil site, no matter how similar to another, has had its own particular genesis, and even though certain natural phenomena may be interlinked and repeat periodically, in detail each natural event is unique. So what happened at Pietraroia, on that ancient island? What are the natural events that turned that outcrop into a *Lagerstätte* that the whole paleontological world envies? In general outline, shortly after their death, the animals at Pietraroia were rapidly buried and protected by a very fine-grained silt that was deposited on the bottom of a lagoon covered by a shallow sea, in an environment with a tropical climate similar to that of the present-day Bahamas. Under normal conditions, with a calm sea and strong sunshine, vertical mixing of the waters was so slow that the bottom of this basin was periodically totally deprived of oxygen. Consequently, decomposition of the dead creatures that were deposited there was brought almost to a halt. Compaction and dehydration of the marine ooze was accompanied by a slow process of replacement of the remains of the organisms, molecule by molecule, with mineral substances composed of microscopic calcite and apatite crystals, which have thus meticulously preserved the shape and structure of the anatomical parts of plants and animals. Progressive silting up of the basin by the sediments, cutting off the Tethys, and subsequent upheaval of the earth's crust about 40 million years ago led to the formation of the present-day Matese Massif, which, like a jewelry box, still holds these precious fossils.

But paleoenvironmental conditions of the lagoon of Pietraroia were a little more complex and have left numerous traces that enable us to reconstruct a much more precise scenario.

Near the township of Pietraroia, the fossil-bearing outcrop, which is still clearly visible in the locality of Le Cavere, is composed of limestone of the Lower Cretaceous period, more or less converted to dolomite rock (i.e., enriched with magnesium) up to 320 meters thick. The richest fossiliferous levels are finely laminated and are contained in the top 50 to 70 meters of the sedimentary series. The age of the Pietraroia limestone has long been a subject of debate by the geologists who have studied it. Recent studies have enabled it to be dated precisely, thanks to more careful observation of the "index fossils."[26] The foraminiferan protozoan *Ovalveolina reicheli* is characteristic of the first part of the Lower Albian, a stage of the Lower Cretaceous period whose absolute age is approximately 110 million years. Even though such a long period of time has passed as to seem inconceivable, it is possible to make quite a detailed description of the environment in which little *Scipionyx* died, thanks to analysis of the sediments (sedimentology) and to the study of the methods of burial and fossilization of the organisms (taphonomy).

The experts thus came to the conclusion that the environment that had enabled "Ciro's" preservation consisted of very shallow coastal lagoons in which sedimentation was due to the sporadic inflow of mud mixed with tiny fragments of organic origin: foraminiferan shells, skeletal remains of sponges (spicules), fragments of seashells. The carbonates were deposited in these basins at a very fast rate, facilitated by particularly intense flood tides and storms, which, by washing and sweeping the surrounding areas above sea level and then reaching the lagoons, poured the finer sediments into them. Silt was deposited on the seabed by turbidity currents and by precipitation, which could reach such intensity as to look like a real underwater "rainstorm."

These events, which were closely linked to the rhythms of the tides and storms, alternated with more or less prolonged pauses in deposition, due in all probability to isolation of the basins from the open sea, which created dystrophic conditions (deriving from oxygen deficiency) in the waters of the lagoon and caused mass annihilation among marine organisms and an excessive production of algae. This was substantially similar to eutrophication, which occurs today especially in closed basins during particularly hot summers. Oxygen was not entirely lacking on the seabed during the deposition of the Pietraroian limestone, as evidenced by the presence of bioturbation, tunnels, and other burrowing traces made in the silt by the passage of invertebrates. However, occasional mass mortality is documented by several levels, which are particularly rich in fishes. On the other hand, the presence of films of algae and bacteria is the most probable origin of the fine laminations of the calcareous layers richest in fossils and apparently had a stabilizing effect on the carcasses: the covering of algae prevented their disarticulation and assisted the creation of a favorable microenvironment for preservation of soft parts.

The occurrence of periods of overheating of the waters in the lagoon of Pietraroia, during which they were evidently very shallow, is proved by the position in which bivalve mollusks fossilized. In fact, certain layers are literally carpeted with bivalves of varying dimensions, which all lie with the valves open upward and still joined at the hinge. The mollusks entered the basin as larvae, as part of the zooplankton, during periods of more frequent exchange with the open sea, and later died there in their masses due to overheating. This phenomenon can be seen in modern lagoons and coastal marshes at northern latitudes and occurs during periods of insufficient circulation when very shallow waters are overheated by strong sunshine.

The exceptional integrity of the vertebrate fossils indicates that the waters of the lagoon were generally calm. Skeletons of fishes and reptiles are almost always perfectly articulated, with even very delicate anatomical parts still connected. Whenever they are disarticulated, there is no preferential direction of alignment of the bones but rather random dispersion, probably due to bioturbation (passage of animals) rather than to currents of water. Conversely, the infrequent disruption of the fish carcasses is due to prolonged postmortem floating, which caused partial disintegration of the carcasses: either the head is missing or the belly is open. These processes are known to occur only at shallow depths and in warm water, because at higher pressures putrefaction gases would not have the force necessary to inflate the abdomens of dead fishes and float them up to the surface.

In any case, "Ciro" suffered a different fate. Several experiments have demonstrated that carcasses that float never reach the bottom intact. Therefore we can definitely say that the little dinosaur was drawn rapidly down to the bottom and was just as quickly buried by the calcareous mud. Compaction of the sediments isolated the interstitial water that saturated the mud from the circulating water that passed from the mud to the seabed. This sequence of events helped create a microenvironment in which oxygen was absent or was present in very limited amounts. Under these conditions, the action of decomposing bacteria is slowed considerably. Thus, there was sufficient time for the anatomical structure of *Scipionyx*'s muscles and internal organs to be completely replaced by mineral microcrystals.

The Pietraroia fossil site abounds with fishes. The hundreds of specimens discovered to date represent almost two dozen species, including sharks, primitive bony fishes, and the ancestors of modern fish. There are also a great variety of marine invertebrates, including a dozen species of mollusks (bivalves, gastropods, ammonites) and three species of decapod crustaceans (shrimps). In addition, even before the dinosaur was discovered, a number of terrestrial lizards, sphenodontids, and semiaquatic reptiles (mesosuchian crocodiles) had been found in the limestone of Pietraroia, as well as rare salamanders and the remains of terrestrial plants (cycads, cycadeoids, and araucarias). These fossils unmistakably bear witness to the existence of areas of dry land with surface fresh water at the periphery of the lagoon.

The absence of evaporite minerals (deposits of salts, similar to those we see in salt lakes, which form readily in the overheated waters of coastal lagoons isolated from the sea) and the presence of freshwater ostracod crustaceans in certain fossil-bearing levels strengthen the hypothesis that there was a supply of river water flowing into the lagoon of Pietraroia, which not only prevented it from becoming overly saline during the periods in which it remained isolated but possibly rendered it brackish or transformed it into a coastal lake.

The terrestrial organisms that are found in association with fishes and marine invertebrates were probably swept into the sea by rivers or during violent hurricanes, but it is difficult to say whether *Scipionyx* drowned or died from other causes. However, now we have more than sufficient information to try to imagine what happened at Pietraroia 110 million years ago. Ours is obviously a hypothetical reconstruction, but the elements and the events discussed are based on scientific data. It may well be that this reconstruction comes very close to what actually happened.

## A Prehistoric Drama

### One summer day, 110 million years ago

The air is still and oppressive. At this latitude, just a few degrees north of the Tropic of Cancer, and in this season, the future Mediterranean is under the full sway of tropical cyclones. On land, the air is laden with humidity, and strong sunshine creates an almost unbearable sultriness. The vegetation is lush, and the forests are teeming with insects, reptiles, and birds; buzzing, hissing, chirping, and twittering animate the dawn of each day, blending into an exotic concert.

On the banks of an estuary, small crocodiles bask lazily in the sun. On the horizon, seen as a fine, continuous blue-green line, extends the Tethys Sea—on all sides, for 360 degrees. Italy is not yet a peninsula as we know it today, but rather an archipelago of islands and islets immersed in a vast shallow ocean. It is now at the center of a subtropical area that is geologically very similar to the present-day Bahamas.

A group of tiny dinosaurs, no more than a meter and a half long, rest among the cycads on a hillock. They are almost all squatting like chickens, with their hind legs tucked under their bodies, but they rest their forelimbs forward with their forearms placed together, like cats. They have three long, slender fingers, which end in the same number of curved talons. Some rest with their necks on the ground, while others keep their necks raised and turn their heads left and right from time to time like a periscope, keeping watch over this year's new hatchlings, which run around noisily nearby. They are bipedal. They have long, slender hind limbs, like little ostriches, and walk with their forelimbs raised high above the ground, sometimes held against their chest with their fingers bent and sometimes extending forward with their fingers wide apart. They jump and change directions with sudden rapid movements, balancing themselves on their long tails. The babies have shorter and stubbier snouts than the older dinosaurs squatting nearby, big round eyes, and sharp little teeth. Judging by their size, they are probably only a few weeks old. They are carnivorous dinosaurs. In body structure, they seem to be more evolved than *Compsognathus* but more primitive than *Velociraptor:* they are a group of *Scipionyx.*

The group is on the edge of a clearing composed of a carpet of ferns, but not in an actual meadow; grass will not appear until 60 million years from now, and flowers, at least those larger than a few millimeters, 10 million years from now. The Lower Cretaceous landscape is still a palette of greens and browns. From the clearing, a wide view can be enjoyed over the lagoon, which, with its extremely fine sandy beaches, is a true earthly paradise. It is a lookout, but above all a precaution: from here the little herd of *Scipionyx* can control the territory and foresee the arrival of danger.

The young ones, as curious as ever, explore the surrounding areas. One strays away from the herd in pursuit of a *Chometokadmon* (a small lizard); another one, out of curiosity, follows it at a distance. The frightened *Chometokadmon* makes a sharp about-face and escapes between the first *Scipionyx*'s legs, but it falls right into the clutches of the second, who slips it into its mouth as quick as a wink. The first little dinosaur refuses to give in and grips the dangling tail of the prey with its teeth. The pair of *Scipionyx* fight over it with such perseverance that after coming away empty-handed, the defeated *Scipionyx* is exhausted.[27]

Meanwhile, beyond the barrier reef the waves surge with an ever-increasing roar. A light breeze suddenly blows up, heralding a change in the weather. Large, high clouds billow on the horizon, becoming increasingly dense and laden with rain. The pale blue sky has now turned black, while the sea takes on a purple tinge. Huge breakers dash and ride over the barrier of rudists,[28] dispersing in the lagoon and then swelling again toward the shore, where they crash violently, eroding and churning up the once perfectly smooth sandy beach. The most violent hurricane of the season is heading straight for the coast. The island is hit full on, violently shaken by the fury of the elements. The herd of *Scipionyx* seeks refuge in the depths of the vegetation and among the rocks. In the general scramble for cover, each dinosaur looks out for itself as best it can, without paying much attention to the others. The young ones that did not stray too far follow several adults step by step and manage to reach safety. But the baby *Scipionyx* defeated in the battle over the *Chometokadmon* is lost in the clearing that slopes down toward the plain and the river mouth. To take shelter from the raging wind, the dinosaur crouches in a dip in the ground, but a torrent of water and mud suddenly engulfs it and sweeps it down to the lagoon, together with cycad branches and leaves. The little dinosaur, desperately struggling against the destructive force of the hurricane, reacts instinctively and hisses, as though to frighten an enemy. Then, at the mercy of the current, it utters a stifled shriek of terror. With its mouth still gaping, it flounders, desperately thrashing its feet and tail, and in a few seconds, choked by the debris and water, it drowns.

"The little dinosaur, desperately struggling against the destructive force of the hurricane, reacts instinctively and hisses, as though to frighten an enemy. Then, at the mercy of the current, it utters a stifled shriek of terror." *Drawing by F. Fogliazza.*

Under the water, in the lagoon, all sounds suddenly diminish. For a few seconds the gurgling of the undertow can still be heard, and then only a smooth swashing. Detritus disperses, and suspended sediment forms clouds of darkness. On the bottom, strong ebb currents drag the tiny body of the unfortunate little *Scipionyx* down the slope of the lagoon toward the center of the basin, together with a large quantity of mud and fine calcareous silt. No predator can take advantage of this tasty, tender morsel. In heavy seas, the sharks and the *Belonostomus,* a voracious predatory fish similar to a moray eel with the snout of a barracuda, prefer to stay offshore, in deeper waters far from the coast.

When the wind and the waters calm down and a rainbow hails the return of peace and quiet on the island, the little dinosaur at the bottom of the lagoon is already covered with a soft blanket of sediment, which in its finer fractions continues to shower down from above like underwater rain. Only a slight mound of sand marks the carcass's position on the deserted seabed.

## Two weeks later

Fifty centimeters of sediment have taken a little dinosaur, whose life lasted only a few weeks, away from its world and the sun-drenched island forever. However, an eternal fate awaits it. As its skin and outer tissues decompose, dissolved acids and phosphates disperse through water in the pore spaces of the surrounding sediment. Some of the internal organs and muscles form a barrier, circumscribing a microenvironment that allows the accumulation of a higher concentration of acids and phosphates sufficient to cause rapid precipitation of the phosphorus and calcium salts responsible for mineralization. Molecule by molecule, organic compounds are replaced by inorganic mineral crystals, which, reutilizing many of the atoms present in the tissues and arranging themselves exactly like the latter, maintain the same structure at a microscopic level that the cells of the animal had. The little *Scipionyx* is already fossilizing.[29]

## Six months later

In full dry season, after a prolonged period of drought, the rivers are reduced to streamlets, and the lagoon, no longer supplied with runoff from the land, drops to such a level as to remain isolated from the open sea. The temperature of the water rises, evaporation increases, and trapped fish are subjected to increasing thermal stress. In the salty pools overcrowded with marine animals, little crocodiles and coastal birds gorge themselves on mollusks, crustaceans, and small fish. With only 2 meters of water in its deepest part, the lagoon is flat and without waves, because the surface of the water is covered by a mucilaginous blanket of seaweed that restrains its movement. On the bottom, worms, ophiuroids, bivalves, and shrimp are already dead. Starting at the bottom, as decomposition products accumulate, oxygen levels become lower and lower. An invisible death from increasing stagnation rises to the surface and envelops the underwater community in front of the tropical island, until anoxia causes widespread death of fish. Small sharks, pycnodonts, and primitive sardines accumulate in the sediment above the little dinosaur, while in its natural grave sediment layers compact, compressing it more and more. Marine life will return to the lagoon in a few weeks' time, after heavy spring rainstorms fill the rivers, helping waters in the lagoon to breach the sandy barrier that isolated the lagoon from the open sea.

## A few thousand years later

With the passing of time, the bones and remains of the soft parts of the little dinosaur, now completely replaced by minerals, are deformed under the slow but increasingly strong pressure of the calcareous substances incorporating it, which have now turned to limestone, and the sediments that continue to accumulate on the bottom of the Tethys basin, forming thick and increasingly solid rock. On the surface, the landscape is changing. The placid lagoon no longer exists. The level of the sea has risen by only a meter, but the island of the *Scipionyx* has grown very much smaller. The area above sea level is reduced by half, causing overcrowding among the terrestrial animals living on it. Competition has increased, food is scarce, and the ecological balance of long ago no longer exists. The vegetation does not have time to replenish itself, and this will sooner or later lead to collapse of the food chain. Survival of the endemic species, especially the larger ones, is in serious question. First the large herbivorous animals will disappear; then the small carnivores will go. Among the ancient inhabitants of the island, the dinosaurs are destined to disappear completely. They will all die of hunger within the space of a few years. Among these are the last of the *Scipionyx*.

## 40 million years ago

The bottom of what was once a tropical sea, now completely filled with sediments compacted into a vast calcareous platform, has begun to compress, break up into small plates, and emerge, held in a vice-like grip, between the Eurasiatic and African continental plates, which are closing in on each other. The Tethys Sea is withdrawing further to both the east and west. The emerging area will become the Apennines. There are no longer any islands or land bridges but rather a restless peninsula, made of quaking earth, bordered here and there by mountains that belch fire, crossed by vast grassy plains inhabited by herds of shaggy animals. The dinosaurs that did not learn to fly (i.e., all the dinosaurs other than birds) have been swept away by a cataclysm of planetary scale. Mammals have inherited the dry lands previously dominated by the reptiles and are exploding into a myriad of forms, some of which are also colonizing the seas and the skies.

### 4 million years ago

Freed by erosion of overlying rocks, the bottom of the lagoon is again emerging into the sunlight. But there is no sea, and there is no warm wind of the tropics. Now the bottom of the lagoon is a white slope of hard, compact limestone. It is cold here and will be even colder in the future, when glaciers descend from the north over all the high lands. We are at an altitude of almost 1,000 meters, on mountains that in a few million years' time men will call the Matese Apennine Massif. The Samnites will wage war here, the village of Pietraroia will be built here, and it is here, in 1798 A.D., that the geologist Scipione Breislak will write about strange fishes in the rock.

Of course, he could not have imagined that just a little farther in the future, the same place would disclose much more than a fish or a petrified shrimp: a dinosaur that would bear his name.

# 4

# "ANTONIO" AND THE HADROSAURS OF THE TRIESTINE KARST

About 85 million years ago, whole herds of dinosaurs roamed in the area corresponding to the present-day Triestine Karst. They belonged to the family of the hadrosaurs, also referred to as duck-billed dinosaurs because of the peculiar shape of their wide, flat snout. One of them was discovered in 1994 and extracted in 1999 during an excavation campaign not far from Trieste, in an abandoned quarry in the borough of Duino Aurisina at Villaggio del Pescatore (Fisherman's Village), 100 meters from the sea.

It all started with a report made by Alceo Tarlao and Giorgio Rimoli in the 1980s. While prospecting for rare minerals, they came across several black fragments outcropping from the gray calcareous rocks, which looked for all the world like bones. They knew very well that those rocks had formed during the Cretaceous Period, and that consequently these could be nothing other than the remains of some large reptile. They selected a bone fragment and compared its microscopic structure with one shown in the photographs of a book. It really was a dinosaur!

"Antonio," as the dinosaur extracted in 1999 was christened by the press, is in an exceptional state of preservation and in perfect articulation. It is an adult specimen 4 meters long, 1.3 meters high at the hip, with an estimated weight in life of approximately 350 kilograms. Studies are only at the initial stage, but "Antonio" promises big surprises. It is the most complete and most ancient hadrosaur ever found in Europe. As far as is presently known, on the basis of the fossils found in France, Spain, the Netherlands, and Romania, these dinosaurs appeared on the European continent about 70 million years ago.

# 100
## Dinosaurs of Italy

According to Fabio Dalla Vecchia, the paleontologist who was scientific director of the excavation appointed by the Italian Ministry of Cultural Heritage, "Antonio" possesses very primitive characteristics, which can give us some important information about the evolutionary background of this extraordinary group of prehistoric animals. In fact, "Antonio" is at least 10 million years older than its more famous and numerous American "cousins" and therefore almost certainly belongs to a species entirely new to science. This is corroborated by the uniqueness of a number of anatomical details, such as the relatively small size of the skeleton, the very large infratemporal fenestrae, a lower leg that is longer than the thigh (which denotes a predisposition to run), very long ischia, and a forefoot with only three fingers. The site of Villaggio del Pescatore promises to yield additional interesting discoveries, because to date only a small part of the site has been excavated.

To extract the dinosaur, the technicians of Stoneage, the company entrusted by the Italian Ministry of Cultural Heritage to carry out the excavation work, had to remove more than 300 cubic meters of rock with the help of a special steel cable with diamond inserts. In the outcrop of Villaggio del Pescatore, the strata lie vertically (so that "Antonio's" skeleton appeared to be "standing up"), so the face of the excavation had to be shifted back to reach the specimens deep inside and, after literally slicing off the rock, to free a sufficiently large area of strata in order to recover them. To cut the blocks, it was necessary to make use of equipment identical to that used in big marble quarries. An auger drilled the eight corners of a cubic portion of rock. A steel cable was then inserted into these channels, closed to form a ring, and then made to slide at high speed by large pulleys driven by a powerful motor until all the sides of the cube of rock were cut.

In addition to the complete hadrosaur, other fossils belonging to about half a dozen dinosaurs of the same type were recovered during the work. A skull and associated jawbone and other bones belonging to a skeleton slightly bigger than "Antonio" had already been extracted a few years before. The excavations yielded the fossil remains of small fishes, shrimps, and rare plants, plus three small crocodiles also representing a new species, characterized by a short, rounded snout. Quite probably, dozens of dinosaur bones are still trapped in the deposit.

Isolating the portion of rock containing "Antonio" was only the first step in a long process of cleaning up the fossil. In fact, the size of the blocks containing the specimen prevented it from being transported

Villaggio del Pescatore (Trieste), April 1999: the difficult operations of removing the block of rock containing "Antonio's" body (*top right*). The vertical position of the stratifications can be seen under the arm of the excavator. *© Italian Ministry for Cultural Heritage and Activities, Superintendence of Friuli Venezia Giulia. Photo by Stoneage.*

whole. The fossil was initially encased in a volume of almost 2 cubic meters of rock, weighing approximately 5 tons. "Antonio" was therefore removed from the quarry in six separate blocks, some of which needed to be further reduced in weight and were whittled down by cutting away the portions that did not contain bones.

At the laboratory, after having ascertained the marked difference in chemical composition between the fossil and the rock (the latter is a calcium carbonate, while the bones are composed of calcium phosphate heavily impregnated with organic material), and in order to free it from the rock in which it was tightly encased, paleontologists opted for a technique using acids capable of etching and dissolving the rock matrix without damaging the dinosaur. This procedure consisted of continuously spraying jets of a highly diluted aqueous solution of formic acid onto the specimen for many hours, using a series of closed-circuit pumps. After each acid treatment, the fossil was rinsed thoroughly with pure water, and the parts already freed were consolidated with special resins. Then

A detail showing the crushed skull, seen in cross-section at the excavation site. *© Italian Ministry for Cultural Heritage and Activities, Superintendence of Friuli Venezia Giulia. Photo by Stoneage.*

the cycle was resumed. This operation took approximately 2,800 hours of work and delayed "Antonio's" official presentation until December 2000.

## Tropical Forests and Swamps . . . of the Adriatic Sea

Fishes and crocodiles, and above all the carcass of the hadrosaur, paint a landscape very different from that of present-day Italy. The discovery of dinosaurs in the Triestine Karst is an event that revolutionized interpretations of the environment and geography of this area as it was 85 million years ago in the Santonian stage of the Upper Cretaceous Period. The landscape could not have been completely submerged by the sea, as was previously hypothesized, because the presence of large terrestrial animals such as dinosaurs bears witness to the existence of vast areas of emergent land with lush vegetation and inland sources of fresh water. And such paleoenvironmental conditions were not short-lived, because the bones of the hadrosaurs at Villaggio del Pescatore occur at several stratigraphic levels within a 10-meter-thick lens. This indicates that the area was frequented for hundreds and perhaps thousands of years. The present excavation site where the hadrosaurs were found seems to have been an ancient small basin (maybe a coastal lagoon) with warm still waters that in successive stages became a "tomb" for different species of animals. According to Fabio Dalla Vecchia, undisturbed laminations and the black color of the limestone containing the fossils indicate that there was very little oxygen on the bottom of the basin (the black color derives from the high percentage of organic matter, which, due to the lack of oxygen, was not completely decomposed). It seems that these rocks continue laterally for no more than about 100 meters. The layer of black limestone decreases in thickness toward the outer edges to form what geologists call a "lens." This means that the basin of deposition was not very extensive. The Karst was probably full of dinosaurs, but those that did not end up in the basin after death did not undergo the particular processes of preservation that enabled them to become fossils.

However, the fossil-bearing outcrop covers an area large enough that intensive excavation could yield hundreds of bones. Their density obviously varies from one layer to another, and this gives us helpful information about how environmental conditions changed over the course of time. For example, in some strata the bones are not anatomically connected but lie scattered. This, together with the lack of lamination of the sediments, indicates a relatively high-energy environment with currents that disarticulated the carcasses of the animals and scattered their bones. In contrast, in another section of the excavation site there is a stratum composed of fine layers of sediment containing the most complete and best-preserved skeletons. These did not undergo dismemberment, which indicates that a basin with calm, stagnant water had formed in the area during that period of time. The remains of dinosaurs reached this basin after brief transportation on the surface and then ended up on the muddy bottom, which incorporated them and maintained their bones in anatomical connection.

Considering the paleoenvironment on a larger scale, it should be remembered that during the Upper Cretaceous, southern Europe was an island surrounded by a vast shallow sea, within a system once again geologically classifiable as a carbonate platform. This platform included part of northeastern Italy, Istria, Dalmatia, and inland Croatia and reached as far as Bulgaria. According to Dalla Vecchia, "Antonio's" ancestors and those of the European hadrosaurs must have come from Asia, because if they had migrated from North America, we should be able to find their remains in Western Europe, not only in rocks of the Maastrichtian stage (65 million years ago) but also in more ancient sediments. Furthermore, no trace of hadrosaurs—not even a single bone or tooth—has ever been found in Africa.

In "Antonio's" days, the Adriatic region must have had a much warmer, more humid subtropical climate that allowed the existence of lush vegetation. Along the coast and farther inland there were forests of tall araucarias of the genus *Cunninghamites* with a growth habit similar to the sequoias and frutescent (i.e., shrubby with a short trunk) conifers; remains of leafy fronds and leaves have been found in nearby sites of the same age as Villaggio del Pescatore.

## Male or Female?

Our hadrosaur, which died from causes that in all probability we will never know, perhaps from starvation or disease (the bones do not reveal any lesions or injuries typical of a violent death), ended up in a brackish swamp temporarily isolated from the sea, where, once deposited on the bottom, it was quickly covered by sediment. Quick burial, together with the characteristics of the lagoon environment, enabled the skeleton to be fossilized. In fact, the particular chemical composition of the water, the lack of dissolved oxygen, and the fine grain size of the sediment not only ensured the total absence of predators and necrophagous organisms on the bottom, they also prevented the complete decomposition of the body, allowing the skeleton to be preserved almost perfectly for millions and millions of years.

In the limestone slab in which he is fossilized, "Antonio" appears to be lying on his left side with his neck thrown backward in the typical postmortem position (due to tension of the cervical tendons following death and dehydration of the carcass), but with the rest of the spinal column almost rectilinear. In contrast to the position seen in many skeletons of carnivorous dinosaurs, the tail is not arched upward like the neck, most probably due to the ossified tendons arranged in lattice fashion along the vertebral spines of the rear part of the body. These tendons, which are typical of all hadrosaurs, served to keep the back and tail rather stiff during life. The tail was an important balancer for these animals, enabling them to walk or run on their hind legs without their head and body tipping forward. It also facilitated elevation of the body around the fulcrum composed of the pelvic girdle, thereby enabling these dinosaurs

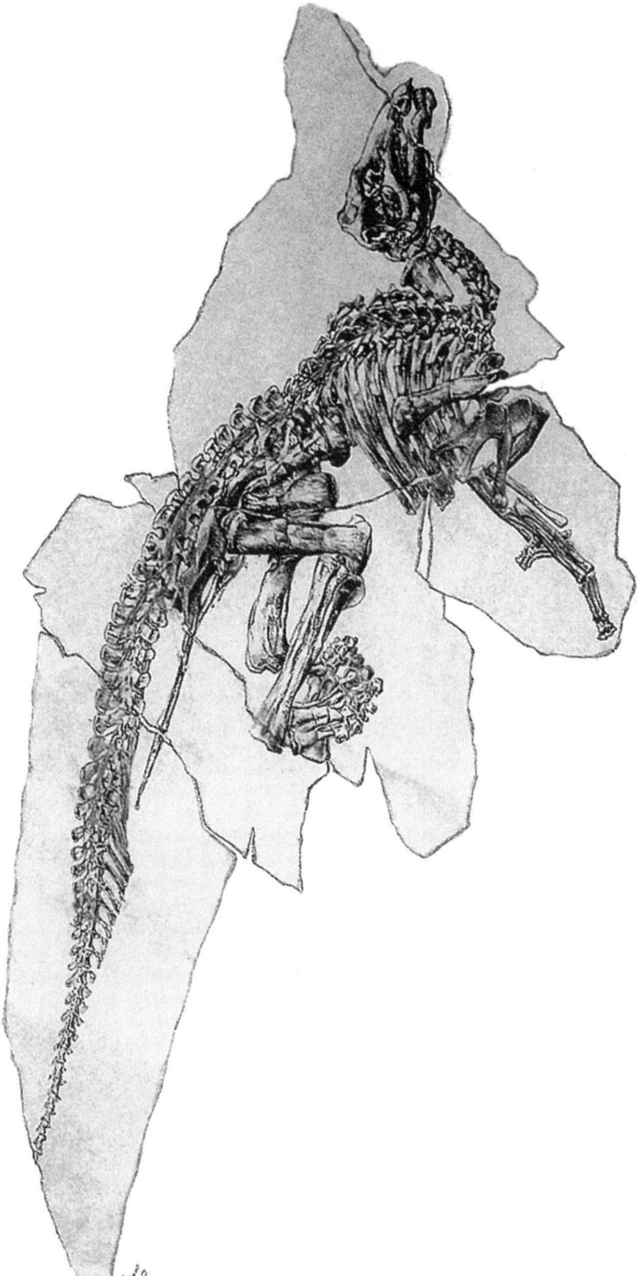

"Antonio's" fossil skeleton as it appears today, after the long work of extraction.
*© Stoneage. Drawing by T. Perentin.*

to lift their heads toward the highest leafy branches of the trees with a minimum of muscular effort. The tail was also an important anchor for muscles that retracted the hind limbs.

From the shape of the head, which lies on its side, it is obvious that "Antonio" had neither crests nor protuberances, and that although this animal is smaller in size, it must have been similar in appearance to the telmatosaurs (hadrosaurs of Romania), with a barely accentuated nasal arch. This leads us to presume, as we will see very shortly, that the Triestine hadrosaurs belonged to the subfamily Hadrosaurinae.

"Antonio's" forelimbs clearly show a vertical position of the metacarpal bones and the proximal phalanges, all grouped to form columns very similar to the extremities of elephants and rhinoceroses: only the terminal phalanges rested on the ground, in a digitigrade posture, which for us humans can be compared to that of classical ballet dancers on tiptoe. The phalanges of the two lateral fingers are more posterior than the central one and form an arch that left a horseshoe-shaped print when the animal walked, with three indentations in correspondence with the tips of the hooves, similar to the footprints fossilized at Altamura.

Analysis of cross-sections of the limb bones revealed the presence of growth rings, the number of which enables us to say that "Antonio" had five or six periods of interruption or slowdown in body growth. According to some paleontologists, this slowing down corresponds to the winter seasons, as in trees. Consequently, the rings appear to indicate that the dinosaur died at the age of 5 or 6. However, other paleontologists are not convinced that a ring corresponds to a solar year, because the climate in the Mesozoic Era was different from that of today, and above all because tests carried out on complete skeletons have revealed that these rings are present in variable numbers in different bones of the same specimen.

Other anatomical details lead us to suspect that "Antonio" was a female. In fact, it was noted that, unlike other hadrosaurs, this specimen has fewer chevrons (V-shaped bones attached beneath the vertebrae) at the base of the tail, and they start much farther back. It has been reported that among present-day crocodiles, females have fewer chevrons than males, and that this difference was observed in several carnivorous dinosaurs such as the tyrannosaurs (a feature that could be linked to the reproductive function, in particular to the need to leave more room for the egg to be expelled from the cloaca). Recently, however, while studying the skeleton of the tyrannosaur "Sue," believed to be a female, Chris

By studying "Antonio's" skeleton, it is possible to reconstruct its functional anatomy, estimate the shape and size of the muscles, and, consequently, imagine the animal as it appeared when alive. © *Stoneage. Drawing by T. Perentin.*

Brochu of the Field Museum of Chicago reexamined the anatomy of crocodiles and discovered that in actual fact, the lack of the initial chevrons reflects intraspecific variability rather than sexual dimorphism. So all these hypotheses have crumbled, and today we are still unable to know whether the most appropriate nickname for the Italian hadrosaur is "Antonio" or "Antonia."

## The Bison of the Mesozoic

As mentioned previously, studies of the dinosaurs of Duino have only just begun, and they have not yet provided all the information that we expect to get from them. We know that these hadrosaurs were small in size and were perhaps even affected by a sort of dwarfism typical of insular species. In fact, the Adriatic hadrosaurs weighed no more than 10 percent of the average weight of an adult hadrosaur of other species. Dwarf dinosaurs—sauropods and hadrosaurs the size of cows—have been found in Romania; *Telmatosaurus* may even be a descendant of "Antonio" and his relatives. (The Romanian hadrosaurs are geologically younger than those of Villaggio del Pescatore.)

There are many things that we do know in general about hadrosaurs that can be useful in giving us a clearer understanding of "Antonio" and his relatives from Villaggio del Pescatore. Hadrosaurs were, in fact, the most numerous, diversified, and widespread herbivorous dinosaurs of the Upper Cretaceous. They grew to a considerable size: adults were usually 7 to 10 meters long and weighed an average of 3 tons, and some species could measure up to 15 meters and more (enormous skeletons have been reported from the Cretaceous sandstone of Mexico).

Duck-billed dinosaurs belong to the suborder of ornithopods, a group of ornithischians that evolved from the iguanodontids approximately 100 million years ago. Their region of origin seems to have been central or eastern Asia, although a North American origin has been hypothesized. Hadrosaurs are subdivided into two subfamilies: those with a flat skull or slightly humped snout (Hadrosaurinae), to which "Antonio" probably belongs, and those with a skull adorned with hollow protuberances of different shapes and sizes (Lambeosaurinae).

Abundant remains of these animals, both complete skeletons and disjointed bones, can be found in the United States (the first hadrosaur was found in New Jersey in 1858) and, especially, in the Canadian province of Alberta. It is on the basis of the discoveries made in North America that paleontologists have been able to reconstruct a sufficiently reliable outline of what hadrosaurs looked like, how they moved, reproduced, and grew, what they fed on, and what they ate. Rare remains of hadrosaurs dating back to 95 million years ago have been discovered in Central and South America and in Asia, whereas in Europe, with the exception of the Italian duckbills, finds prior to the Maastrichtian stage (approximately 71 million years ago) are not known. Hadrosaurs have never been found in Africa, Madagascar, India, or Australia.

Given their lifestyle and paleoecology, hadrosaurs can be considered the bison of prehistoric times. They moved in large herds, perhaps hierarchically organized, and must have been constantly on the move, spending most of their days searching for the plants they ate. We know that while they were able to walk on their hind legs, using their tail to balance, they usually moved on four legs. Their foot was composed of three stumpy toes and differed from the tridactyl foot of carnivorous dinosaurs in lacking talons but having characteristic little hooves. They lived on coastal plains, located close to the sea or near swamps, alluvial plains, and wide meandering rivers, where the vegetation was more plentiful.

For a long time, it was thought that hadrosaurs even led an amphibious life. This hypothesis was based on an erroneous interpretation of certain anatomical features, such as the laterally flat, broad tail, which seemed particularly suitable for swimming, and the fingers of the hands, which in several "mummified" specimens seemed to be webbed like the feet of ducks, but in fact were united in a fleshy pad that served to cushion their weight on the ground.

Hadrosaurs were among the few dinosaurs capable of really chewing their food. And they did so in a unique way, which has no equivalent in the whole animal kingdom. The first half of the mouth, externally distinguished by the characteristic "beak," was completely devoid of teeth. They gripped the leafy branches of plants by closing the beak and then simply tearing them away, as many modern herbivorous mammals do with their lips. In the posterior portion of their mouths, hadrosaurs possessed complex and efficient "dental batteries" composed of innumerable compact rows of tiny teeth (more than 1,000 in the adults), which formed a highly efficient grinding surface.

Hadrosaurs were the only terrestrial vertebrates with movable upper jaws. Their maxillary bones were not completely united with the rest of the skull but had special joints, so that when the jaw closed, the lower dental batteries pressed against the upper ones and caused them to widen out on both sides, toward the outside of the mouth. Thus the teeth did not simply close against each other but "rubbed" sideways, grinding food like a mill. The teeth were replaced continuously due to wear, which enabled even old individuals to chew tough, fibrous plants.

## Loving Parents

A number of exceptional discoveries have enabled us to take a real journey back in time to discover many aspects related to the social be-

Reconstruction of the environment of Villaggio del Pescatore as it must have been about 80 million years ago in the Upper Cretaceous Period. The hadrosaurs grazed near a brackish pool not far from the sea, which was also inhabited by crocodiles and turtles. *© Stoneage. Drawing by T. Perentin.*

havior of these docile animals. We have proof, for example, that hadrosaurs cared for their young.

In 1979, American paleontologists Jack Horner and Bob Makela, while excavating an Upper Cretaceous site in Montana, exposed an entire area that appeared to have been used as a nesting place by a group of hadrosaurs belonging to a new species, which Horner himself named *Maiasaura* (or "good mother lizard") to highlight what he felt was its most important characteristic: care of its young. After finding the remains of a cone-shaped nest of mud, almost 1 meter deep at the center and 2 meters in diameter, which contained the bodies of fifteen fossilized baby dinosaurs, Horner concluded that these little animals had remained in the nest for a certain period. Their skeletons were poorly ossified but quite developed, and above all, their teeth showed signs of wear due to chewing. Consequently, they had been born some time before, and the food had been brought to them by the adults.[1]

At the end of the excavations, the two researchers collected hundreds of eggs and dozens of skeletons of dinosaurs of different ages. The nests were spaced no less than 7 meters apart, the approximate length of an adult hadrosaur. They deduced from all this that they had come across a nesting site, and that the *Maiasaura* (and perhaps also other members of the subfamily Hadrosaurinae) lived in a group, watched over the nests, and continued to take care of their young for a long time after the eggs had hatched, a bit like what occurs today in certain seabird colonies.

In 1987, in a locality known as Devil's Coulee in Alberta, Philip J. Currie and a team from the Royal Tyrrell Museum in Drumheller discovered nests, eggs, and skeletons of newly hatched *Hypacrosaurus* (a kind of hadrosaur belonging to the second subfamily, the Lambeosaurinae), which revealed a reproductive behavioral pattern similar to that of *Maiasaura*. In all probability, the ability to take care of their young was a feature common to all the duck-billed dinosaurs, and this gives us reason to believe that the dinosaurs of Villaggio del Pescatore behaved likewise.

## A Sound from the Past

We know a great deal about the appearance of hadrosaurs. We know, for example, what their skin was like. Skin is very rarely preserved, but thanks to a number of exceptional discoveries, it has been possible to obtain some information about this feature. The greatest contribution was made by the two famous hadrosaur "mummies" of Wyoming, now housed in the American Museum of Natural History of New York and in the Senckenberg Museum in Frankfurt. The first was discovered in 1908 by the famous fossil hunter Charles Sternberg and his three sons. The group was busy working at an excavation site near the town of Lusk in southern Wyoming, exposing the remains of a duck-billed dinosaur. While the father and one of the boys were in town getting provisions, the other two carried on with the work. After removing a large piece of sandstone from the animal's chest, they suddenly saw a perfect, well-preserved impression of the skin and immediately realized that what lay before them was an extraordinary discovery. On his return, Charles Sternberg carefully analyzed the specimen and noted traces of fossilized skin everywhere. What was it that had enabled the dinosaur's skin to remain preserved?

Sixty-five million years earlier, immediately after the hadrosaur died, the very arid climate, windy weather, and several other particular circumstances had mummified its body, protecting it against decomposition.[2] For some reason of which we are unaware, it had also escaped attack by animals that feed on dead carcasses, ending up by being quickly covered by sand. The sediment then consolidated around the carcass, preserving it just as Sternberg and his sons found it.

The second mummy was discovered in 1910, again in Wyoming and again by the Sternberg family. It is preserved in a similar way, not crushed in a thin layer of sediment but fossilized in three dimensions, and shows not only a nearly complete skeleton and the very rare preserved impression of the skin but also the horny beak.

A more recent, and equally spectacular, hadrosaur with skin impressions is "Leonardo," found in Montana in 2001. The fossilized skin of all these dinosaurs has a granulose, tubercular texture, with small, smooth, slippery nodules. It is similar to the skin of some lizards and shows no trace whatsoever of hairs or feathers.

Even though we do not know this for certain, it is reasonable to believe that these animals were colored. Modern reptiles and birds (the most direct descendants of dinosaurs) have multicolored markings that serve numerous functions: to attract a mate during courtship, to identify them as a member of a particular species, to camouflage them, or to inspire fear.

Some hadrosaurs also sported bizarre cranial crests and protuberances, the purpose of which remains an enigma. In the past, paleontologists who thought that hadrosaurs were semi-aquatic dinosaurs maintained that the crests stored an air supply for when the dinosaurs submerged. In fact, since the crests were probably different in males and in females, they were most likely features linked to courtship. The shape and size of the crests changed according to age: crests were absent in hatchlings, while in young individuals they were not fully developed.

The theory that these structures served to amplify sounds has recently gained considerable support. The most bewildering of these crests was that of *Parasaurolophus,* a hadrosaur weighing 2 or 3 tons that lived in North America during the Late Cretaceous. This crest was more than a meter long and contained a labyrinth of internal chambers for passage of air from the outer nostrils to the lungs. Experiments carried out on complete skulls have enabled researchers, among them David Weishampel of Johns Hopkins University, to re-create the probable call that these animals could emit. Tom Williamson, curator of the New Mexico Museum of Natural History and Science, recently made a computerized three-dimensional reconstruction of the internal structure of the skull, which was then used for simulating the probable sounds produced by *Parasaurolophus.* And the result? A very low, prolonged sound, like that of a horn. Expressed in more poetic terms, it sounds like a giant clearing his throat. After hearing it, some experts even hypothesized that it could have been so distinctive as to enable dinosaurs of the same species to maintain contact with one another at considerable distances, and even in the maze of a forest.

# 5

# JURASSIC, ITALY: THE SALTRIOSAUR

IT WAS A FEROCIOUS predator, one of the largest of the Jurassic Period. And it is the first big carnivorous dinosaur whose skeletal remains have come to light in Italy: 8 meters in length, almost 4 meters in height, and weighing half a ton. A 70-centimeter skull furnished with teeth as sharp as daggers and forelimbs equipped with three powerful fingers armed with talons made it a fearsome war machine, similar in all respects to the North American allosaurs. All that remains of the saltriosaur, as it was named, are a few bones, just enough to demonstrate that this description is not a figment of the imagination. The rest of the skeleton was irremediably lost before the contemporary world became aware of its existence one warm summer morning in 1996. This chance discovery was destined to begin a new, important chapter in the history of Italian dinosaurs.

## The Discovery

Just after five in the afternoon on August 4, 1996, paleontologist Giorgio Teruzzi received an unexpected telephone call at his home. On the line was Angelo Zanella, a member of the Brianteo Geological Research Group of Paina di Giussano (Milan), who told Teruzzi that while looking for fossil shells that morning in a quarry in the province of Varese, he had discovered a boulder that seemed to contain bones. "I didn't collect them because the block of rock was too heavy," he told Teruzzi. "But if they are of any interest, tomorrow I'll try to extract them."

The next morning, upon returning to the quarry with his two young sons, Zanella found the entrance open and not a soul in sight. Zanella had a generator and a powerful drill in the trunk of his car, with which he managed to make holes in the fossil-bearing boulder and insert expanders to divide the block and rid it of the portion without bones. He and his sons then gathered everything they could and returned home with their mysterious load. In the afternoon Teruzzi, puzzled by the telephone call, went to Zanella's home in the small town of Mariano Comense. Opening the car's trunk, Zanella showed him several coffee-colored limestone blocks containing brownish-yellow inclusions, clearly identifiable as bones due to their porous appearance. One of them, divided in half longitudinally and revealing a wide marrow duct, measured more than 30 centimeters in length. Teruzzi immediately ruled out the possibility that the remains could have belonged to a crocodile or an aquatic reptile: "No, the bones of marine reptiles don't have these cavities. Sure, this is a marine sediment, but as you are well aware, it contains fossil invertebrates typical of the Sinemurian stage, and so it definitely belongs to the Jurassic Period. . . . In my opinion, this is a dinosaur."

The unexpected declaration was enough to convince Zanella to hand over the three large blocks of rock to the paleontologist from the Natural History Museum of Milan and to tell him the approximate spot where the discovery was made in the Salnova quarry in Saltrio (Varese), a few dozen meters from the Swiss border. Concerned that there might be other remains, the next day Teruzzi organized an impromptu expedition to the privately owned quarry with a few members of the Paleontological Team of Besano, who were provided with all-terrain vehicles and furnished with equipment normally used in the nearby Triassic fossil excavation sites. The volunteers found the spot indicated by Zanella on a demolished slope, but the search proved difficult because the limestone bank containing the huge reptile had been blown up with dynamite into hundreds of large blocks, which were shortly to be removed by powerful excavators and then transported to the crusher, where what remained of the prehistoric creature would be reduced to gravel for making roadbeds.[1] For five hours, the team clambered up the rubble pile created by the explosion searching for outcropping bones, examining the rocky boulders one by one over a range of several hundred meters. They salvaged other blocks containing isolated bone fragments, the remains of the trunk of a plant, a shark's tooth, nautiloids, and brachiopods. The owner of the quarry was asked to inform the Natural History Museum of Milan should other bones be exposed.

August 1996. Retrieving the remains of the Saltrio dinosaur was difficult because the bones were contained in a massive limestone layer that had already been demolished by dynamite and by the local quarry workers. *Photo by G. Pasini.*

Since he had heard nothing further, the following weekend Teruzzi went to make an on-site check, but nothing remained of the heap of blocks and residual wall. The area had changed so drastically that he was not even able to find the precise spot where seven days earlier the bones had been recovered. However, from a geological examination of the terrain, it was clear that the rock levels containing the bones of the reptile, rich in carbonates of marine origin, were intercalated with muddy layers rich in clays of continental origin that had surfaces with ripple marks (caused by wave action in shallow water) and mud cracks (fractures caused by the drying effects of the sun)—unmistakable and rare signs of episodes of emersion of some land. So it was not surprising that a terrestrial reptile typical of the Jurassic Period had come to light in this area.

The slabs of rock retrieved from the quarry in Saltrio remained stored, but not forgotten, in the basement of the Natural History Museum of Milan for a couple of years. Work in progress on other fossils forced the museum's laboratory to postpone the job of extracting the bones. Finally, in June 1999, after a successful etching test had been carried out, it was decided to use a chemical preparation technique to extract the bones. The rock matrix, which was composed of dolomitized carbonates (carbonites recrystallized into calcium and magnesium minerals), was too hard and compact to be chiseled without damaging the bones. So it was dissolved with formic acid, a weak organic acid that did not etch into the bones, which were protected against corrosion by being periodically impregnated with special solidifiers. It was a very slow process in which a continuous acidizing-washing-drying-impregnating cycle had to be repeated every two days, for a maximum of eight consecutive hours of immersion, in an aqueous solution containing only 5 percent acid. And at least half a ton of rock had to be dissolved in this way.

The extraction of all the bones was completed in May 2000, after 1,800 hours of acid baths. The limestone blocks yielded 119 bone pieces. Although half of them were fragmentary and indeterminable, twenty-one rib segments and twenty more diagnostic bones were identified as parts of the limbs of a large terrestrial reptile. This conclusion confirmed a hypothesis that until then had appeared to be pure fantasy. There was no doubt that the bones were the remains of a dinosaur, a carnivore of enormous dimensions, probably transported by water after its death and fossilized more than 200 million years ago on a shoal of the Jurassic sea of Saltrio.

The discovery was made public during a special press conference held

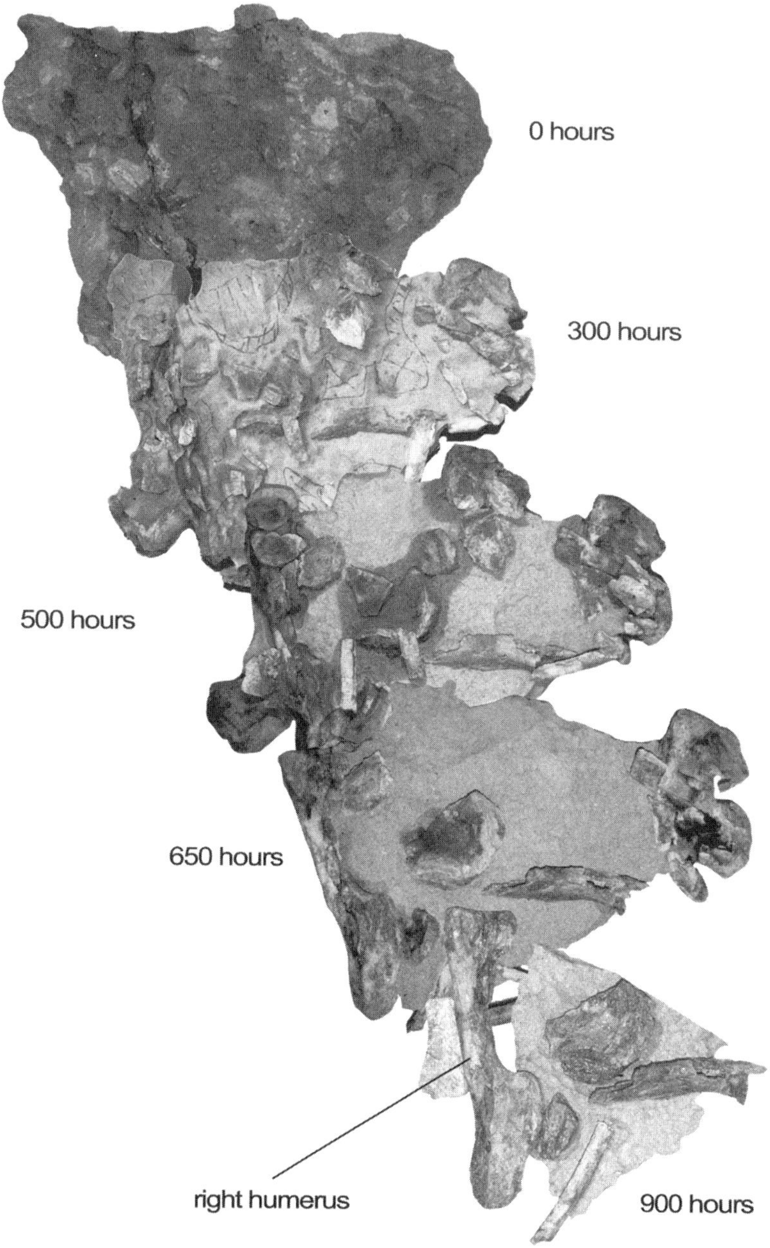


The bones of the dinosaur gradually emerge from one of the slabs of rock recovered in the Saltrio quarry. It took a total of 1,800 hours to dissolve the calcareous matrix in acid and extract all the bones. *© Natural History Museum of Milan, by courtesy of the Italian Ministry for Cultural Heritage and Activities, Superintendence of Lombardy. Photo by C. Dal Sasso.*

on November 9, 2000. At that time, the Saltrio dinosaur was yet to be studied, so on that occasion a brief outline was given of the results of the preliminary analyses.

## An 8-Meter-Long Predator

The Saltrio dinosaur is undoubtedly the first big Italian carnivorous dinosaur. Even though only a few bones have been found, there are sufficient data to say that it represents a new species of theropod. It has been given the provisional name of "saltriosaur"; this name will not be scientifically valid until a precise diagnosis has been made of the animal.

After little *Scipionyx* and the hadrosaurs of Villaggio del Pescatore, the saltriosaur is the third species of dinosaur whose skeletal remains have been found in Italy. At present, it is the first and only dinosaur discovered in Lombardy. Its discovery is extremely important from a scientific standpoint because this specimen comes from rather ancient rocks, dating back to an interval, the Lower Jurassic, during which it was thought that among the carnivorous dinosaurs only more primitive animals called coelophysoids existed.

All carnivorous dinosaurs belong to the order Saurischia and to the suborder Theropoda, and all were bipedal animals. From the very beginning of their evolution, theropods reduced their forelimbs, which were no longer used for walking, and modified them to capture prey. Toward the end of the Triassic and the beginning of the Jurassic, some 210 million years ago, the dominant carnivorous dinosaurs were the coelophysoids, which were provided with four-fingered forelimbs. During the Jurassic Period more highly evolved theropods appeared, called ceratosaurs, still with four fingers on their hand, and another lineage called Tetanurae (literally meaning "stiff-tailed") appeared that reduced the number of fingers to three. From the Jurassic to the Cretaceous, the ceratosaurs continued to spread throughout Gondwana, and the tetanurans superseded them in the rest of the world, dividing into two main branches: the carnosaurs and the coelurosaurs. While the latter led to the emergence of birds, the carnosaurs continued to increase in size and became specialized solely as terrestrial predators.

The most famous representative of the family of carnosaurs is undoubtedly the North American allosaur (*Allosaurus fragilis*). Discovered toward the end of the nineteenth century in the well-known Morrison Formation in the western United States, and initially called by other

The saltriosaur was a carnivorous dinosaur measuring about 8 meters in length and provided with three-fingered forelimbs. Its head must have been rather large and armed with sharp teeth. When it walked it was counterbalanced by a long, thick tail, which was kept raised parallel to the ground. *Drawing by F. Fogliazza.*

names that later fell into synonymy (*Epanterias, Antrodemus*), the allosaur soon became famous due to the wealth of its fossil remains, which enabled complete skeletons to be reconstructed. These reconstructions became the centerpieces of museums in the industrial cities of the East (Washington, New York, Pittsburgh, and New Haven). To convey an idea of the abundance of dinosaur fossils in the Morrison Formation, no fewer than forty-four allosaur skeletons were found in a single outcrop (the Cleveland-Lloyd Quarry in Utah). Younger individuals in this assemblage were 3 to 5 meters long, while several adults may have measured as much as 11 to 12 meters.

Contrary to what was thought until a few years ago, the gigantic tyrannosaurs, which were characterized by enormous heads and only two fingers on the forelimbs, descended not from carnosaurs (nor consequently from the saltriosaur, at least not directly) but from agile little coelurosaurs similar to *Compsognathus.* The Cretaceous tyrannosaurs' apparent similarity to the Jurassic carnosaurs, such as the allosaur, is due to a phenomenon referred to as evolutionary convergence.

During the Cretaceous Period, carnosaurs were not completely replaced by coelurosaurs but continued to spread, especially throughout the continents of the Southern Hemisphere, where in certain cases they attained and exceeded the dimensions of the tyrannosaurids that reigned supreme in the northern continents. Discovered in Patagonia in 1995, *Giganotosaurus,* a distant relative of the allosaurs, had a skull approximately 180 centimeters long and must have measured 13 meters in length; it weighed an estimated 8 tons.[2]

The fact that the saltriosaur comes from rocks that, as their index fossils show, date back to the Sinemurian stage (Lower Jurassic) is extremely important in enabling us to understand the evolution of carnivorous dinosaurs. The first Lombard dinosaur may hold another record as well, because, with the possible exception of the Chinese therizinosaur *Eshanosaurus,* which was described in 2001, it could be the oldest tetanuran in the world. In simpler terms, the Saltrio dinosaur is proof that large theropods with three-fingered hands, the ancestors of the allosaurs, existed 200 million years ago, about 20 million years before they were thought to have originated. In fact, the oldest representative of this group of which we have fossil evidence is the enigmatic *Cryolophosaurus* ("frozen crested reptile"), a dinosaur that resembled an allosaur but was distinguished by a bizarre forward-facing crest on top of the head. As its name suggests, *Cryolophosaurus* comes from one of the most remote and

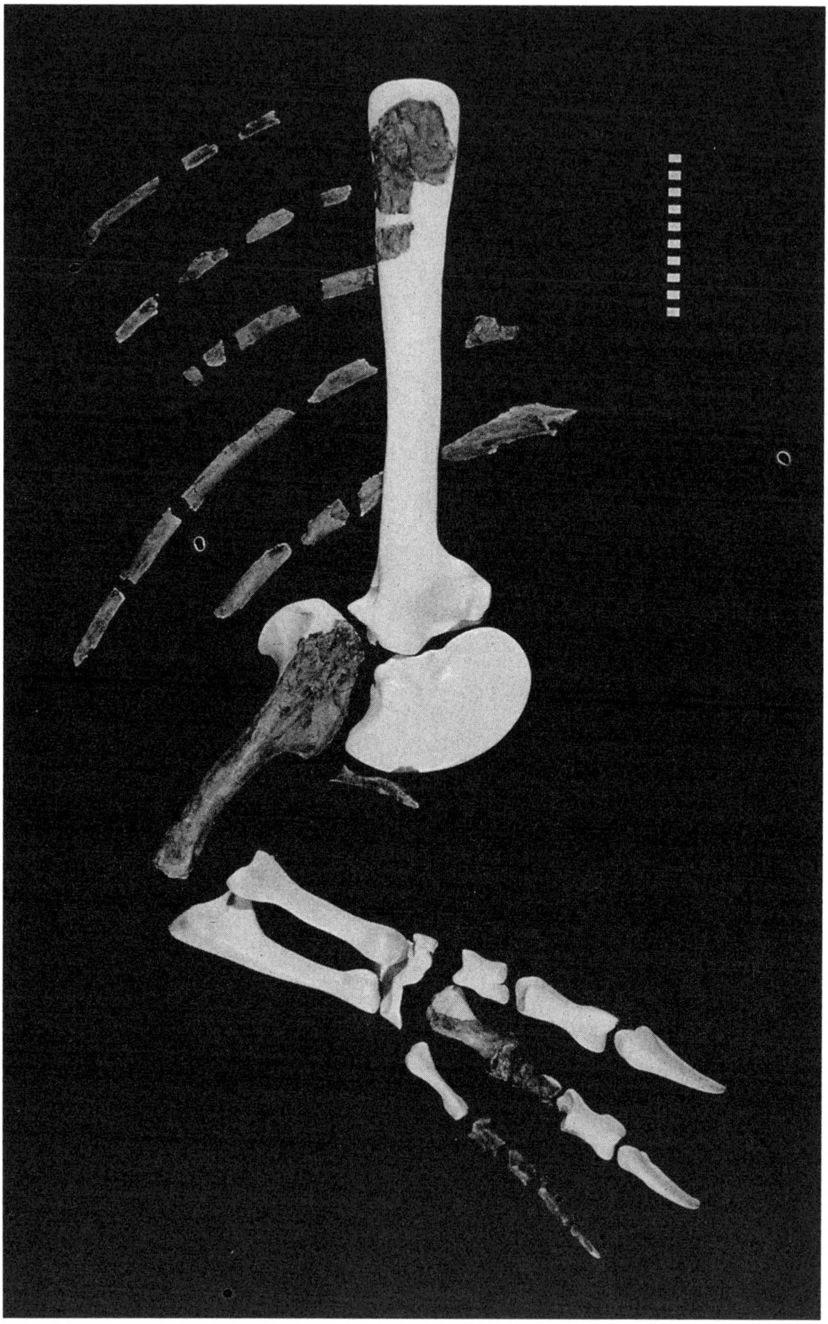

Part of the preserved bones of the Saltrio dinosaur (right forelimb and ribs) recomposed in anatomical position. The missing parts are reconstructed in white; the scale bar measures 20 centimeters. © *Natural History Museum of Milan, by courtesy of the Italian Ministry for Cultural Heritage and Activities, Superintendence of Lombardy. Photo by L. Spezia.*

unlikely areas of the planet for a reptile: its remains were discovered in Antarctica in 1994 by an American expedition and were extracted from rocks that had formed during the Pliensbachian—that is, at a geologic time subsequent to the Sinemurian.

The skeleton extracted from the Antarctic rocks is somewhat incomplete, but the paleontologists who studied it observed primitive features as well as a number of advanced characters which suggest that carnosaurs were differentiated from other tetanurans as early as the Lower Jurassic. According to some experts, the cryolophosaur is actually an allosauroid, a type of dinosaur that, due to its degree of evolution and stratigraphic distribution, is believed to be typical of the Middle-Upper Jurassic. Similarly, the saltriosaur may also turn out to be a primitive allosaur, in which case it would also be the most ancient allosauroid on the planet.

The existence in Europe of large predatory dinosaurs related to the American allosaurs was established in 1978, when the first and most complete skeleton was discovered on the Isle of Wight in England. The specimen, called *Neovenator,* was not studied until 1996. The fossil dates back to the Lower Cretaceous and therefore probably belongs to a more evolved allosaur. It was 7.5 meters long and more slender than the American allosaurs.

Comparison of the preserved bones of the saltriosaur with those of a complete skeleton of *Allosaurus fragilis* displayed in the Museum of Natural Sciences of Bergamo showed that the dimensions of the arm bone (humerus), hand bone (metacarpal), and phalanges of the fingers are identical. The skeleton assembled in Bergamo is 8.7 meters long and has a skull measuring 70 centimeters. By contrast, comparison of the same bones of an allosaur, called *Sinraptor,* found in China in 1993 shows that this animal had slightly longer forelimbs than the theropod of Saltrio but was of a more slender build. In fact, its reconstructed skeleton measures just over 7 meters. Although carnosaurs had a certain variability in body proportions, it seems clear that the saltriosaur was a predatory dinosaur approximately 8 meters in length that could weigh up to 1,500 kilos.

## Not Many Bones but Good Ones

Just under 10 percent of the saltriosaur skeleton was recovered. The most intact bones belong to the right forelimb; there is the end portion of the scapula, a complete humerus, a metacarpal, the first and part of the second phalanx of the second finger, and all four phalanges making

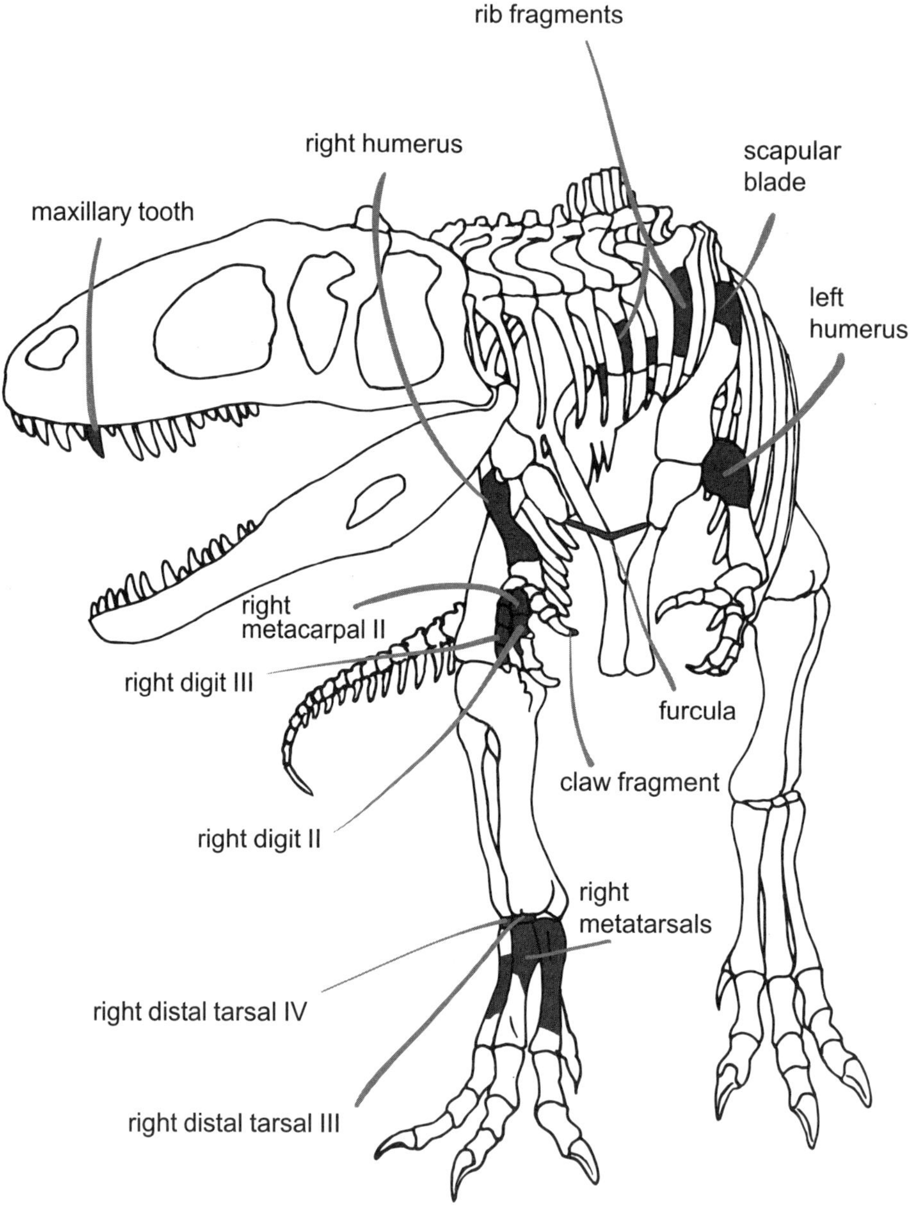


Less than 10 percent of the skeleton (in gray) of the saltriosaur has been recovered. However, twenty-one bones were sufficient to make a reliable diagnosis and for consequent reconstruction of the structure of the entire animal. *© Natural History Museum of Milan. Drawing by F. Nodo and C. Pagliarin.*

up the third finger. The bones are very well preserved; they are three-dimensional and only very slightly deformed, and they retain all the details of the original structure: the spongy tissue inside, the roughness of the muscle insertions on the bone surface, and distinct vascular foramina and articular surfaces. There are also portions of the right hind limb: two ankle bones, called distal tarsal bones III and IV, and three bones of the upper part of the foot, called metatarsals II, III, and IV. Unfortunately, nothing remains of the skull except for an isolated tooth.

The presence of several anatomical characters exclusive to theropods enabled the saltriosaur to be definitively identified as a carnivorous dinosaur. These include laterally flattened, sharp teeth with serrated edges; hollow limb bones; a flat, blade-shaped scapula; a sigmoid (S-shaped) humerus with a prominent muscle attachment called the deltopectoral crest; an elongated hand with outer fingers shorter than the central finger; curved, claw-shaped terminal phalanges; metacarpal bones with deep, clearly visible fossettes on the back of the distal joints; and elongated penultimate phalanges of the hand.

What more can we "read" in the saltriosaur bones?

The forelimbs are very similar to those of the North American allosaurs. The hind limbs reveal advanced anatomical characters in the metatarsal bones, which appear to be, as inferred from their lateral flat surfaces, in close contact with one another, and in the tarsal bones, which are structured so as to enable the ankle to bend in such a way as to prevent the feet from slipping sideways. Moreover, in the pectoral girdle there is a V-shaped bone called the furcula.[3]

The preservation of the furcula is an extremely interesting occurrence—and a true rarity. Its distribution among theropod dinosaurs is much more widespread than was thought until a few years ago: the "wishbone" is not exclusive to birds and their closest ancestors (the coelurosaurs), but has also been identified in the allosaurs, and recently also in the Cretaceous spinosaurid *Suchomimus* and in several Triassic (*Coelophysis*, "*Syntarsus*") and Jurassic (*Segisaurus*) coelophysoids.

Paleontologists are accustomed to shifting the appearance of characters previously believed to be "advanced" back in time as new discoveries lead to greater knowledge of organisms of the past. This is due to the incompleteness of the fossil record. Our understanding of the history of evolution does not form an even slope reaching backward in time, but rather a jagged staircase made up of deep steps with many voids. The reason is that fossils are preserved randomly and rarely, and only during

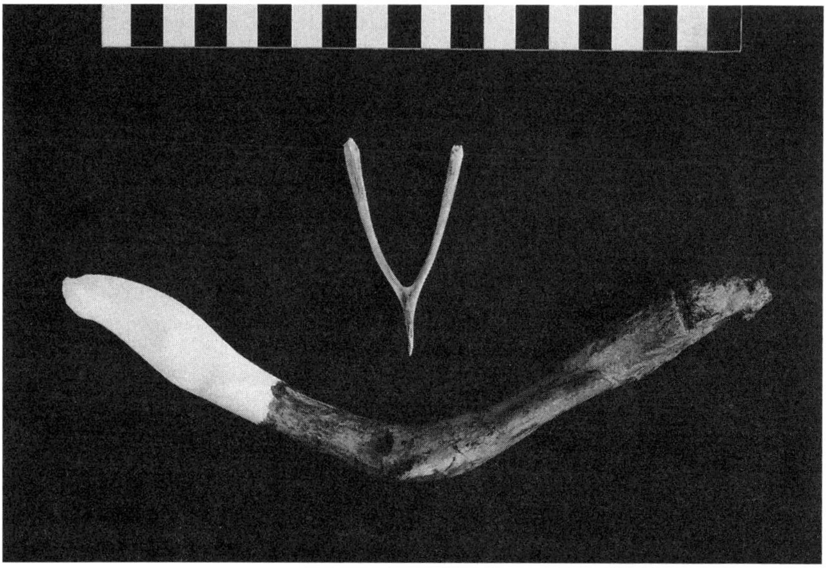

Among the remains of the saltriosaur is a bone that is very rarely preserved in dinosaurs. This is the furcula, commonly called the wishbone (compared here with one from a chicken). The scale bar measures 20 centimeters. © *Natural History Museum of Milan, by courtesy of the Italian Ministry for Cultural Heritage and Activities, Superintendence of Lombardy. Photo by L. Spezia.*

brief periods in which, in certain places and not in others, favorable conditions for preservation of organic remains occur. The case of the megalosaur is a classic example of errors in knowledge due to incompleteness of the fossil record.

Although the genus *Megalosaurus,* described by Reverend William Buckland in 1824, was the first carnivorous dinosaur acknowledged in the history of paleontology, the identity of this animal is still somewhat uncertain. In fact, for almost two centuries, paleontologists continued to describe as megalosaurs many incomplete fossil remains of theropods of very different ages (ranging from the Lower Jurassic to the Upper Cretaceous, an interval of no less than 130 million years!) which came to light in Europe and America. The result was a great deal of confusion. Thanks to the recent discovery of the most complete megalosaur skeleton ever found, we are finally able to understand the megalosaurs more clearly. According to the French paleontologist Ronan Allain, who is currently revising this group, megalosaurs were theropods that lived exclusively during the Middle Jurassic and were less primitive than was initially thought.

In an investigation, we proceed by trial and error, formulating various hypotheses and testing them one by one against the evidence we have at our disposal. We finally find the right road by excluding the other hypotheses. This is also the approach adopted in research aimed at reconstructing the identity of the saltriosaur and its relationships with other theropod dinosaurs.

Since coelophysoids were until now the only carnivorous dinosaurs known to be widespread in the Upper Triassic and Lower Jurassic, they would seem at first glance to be the most probable relatives of the saltriosaur. However, coelophysoids are smaller in size and have a four-fingered hand, while, even though incomplete, the bones of the hand of the dinosaur discovered in Lombardy are much larger and are shaped in such a way as to indicate the presence of only three fingers. One of the "innovative" anatomical characteristics known in the tetanurans is a reduction in size of the third finger compared to the first and second. Comparing the dimensions of the preserved phalanges, this would appear to be precisely the case of the saltriosaur.

The phalanges of the third finger, in particular, are very similar to those of the American carnosaurs belonging to the genus *Allosaurus*. The few differences are in the terminal phalanx (the bony talon), which in the saltriosaur has a much smaller attachment for a tendon, called the flexor tubercle. The second metacarpal bone, with its hourglass shape (with widened ends and a narrow central portion), is also decidedly similar to that of the allosaur but differs in possessing a crest that acts as a stop for the joint of the first phalanx.

On the other hand, the bone of the forelimb (humerus) is sturdier and is not twisted in the form of an S as it is in allosaurs. The humerus resembles that of a megalosaur rather than that of a coelophysoid or a ceratosaur because of the presence of a muscle attachment (the deltopectoral crest) which, in contrast to coelophysoids and ceratosaurs, does not extend halfway along the bone.

Important information about the degree of evolution of the Saltrio dinosaur can be obtained by reconstructing the mobility of the hand joints, especially the movements of flexing and extending. Because the Italian theropod was capable of only a moderate degree of extension, understood as the maximum upward curving of the fingers, it most prob-

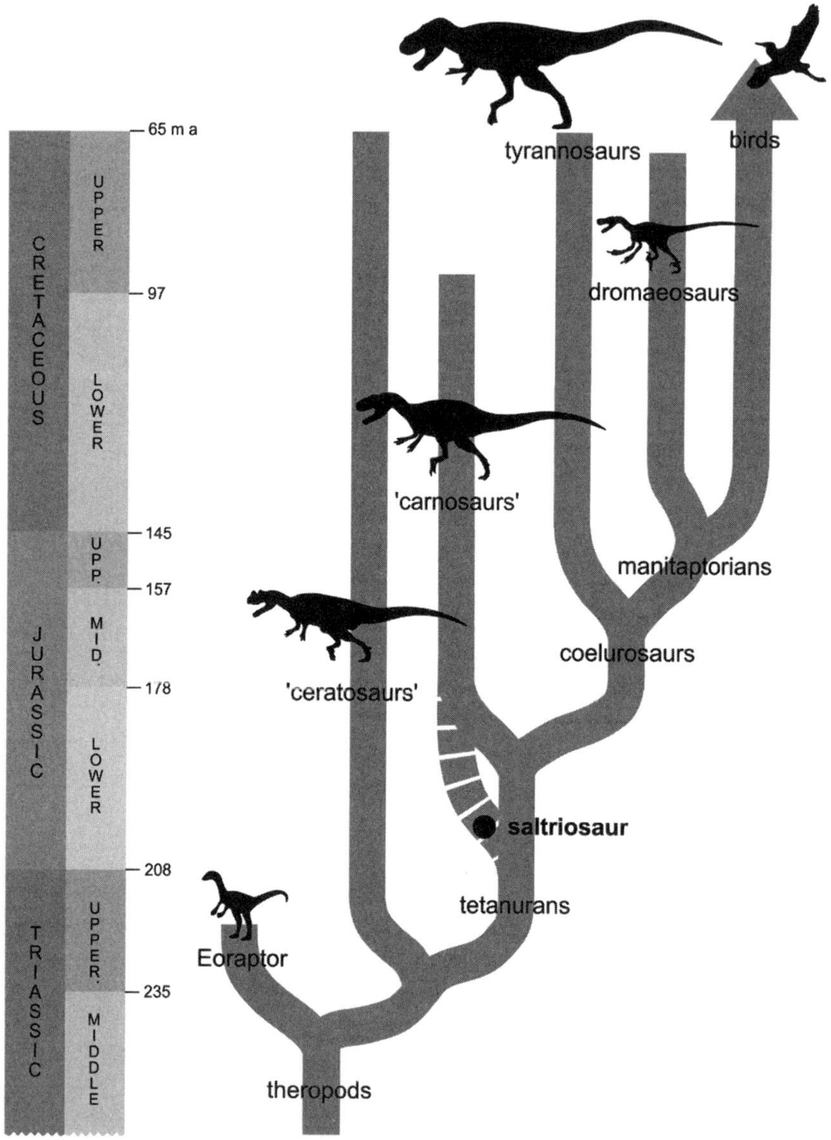


Simplified evolutionary diagram of carnivorous dinosaurs (theropods). The saltriosaur could be the world's most ancient tetanuran and belong to the evolutionary line of the carnosaurs. These dinosaurs, ancestors of the allosaurs, appeared at least 20 million years before they were thought to have appeared (broken line). © *Natural History Museum of Milan. Graphics by C. Pagliarin.*

ably evolved to a grade similar to that of the tetanurans. In contrast, coelophysoids and even more primitive carnivorous dinosaurs were capable of hyperextending the phalanges to almost describe an arch on the back of the hand. This ability was inherited from their distant quadrupedal ancestors, which had to perform these movements when they walked with their forefeet resting on the ground.

Undoubtedly, one of the most fascinating phases in the reconstruction of an extinct animal is studying its functional morphology. Holding the petrified phalanges, putting the articulations together, and trying out the joints are actions that suddenly bring those cold remains to life and make you think of the movements that those hands made some 200 million years ago as the animal captured, grasped, and manipulated its prey.

This sort of "experimental paleontology" is extremely effective in demonstrating the function of an extinct animal's weapons, such as the teeth of carnivorous dinosaurs. Tests carried out on fresh meat with the fossil teeth of *Tyrannosaurus rex,* for example, revealed the secret of their perfect cutting power. Thanks to the dense serrations running along the sharp edges of its teeth, the tyrannosaur's mouth was equipped with

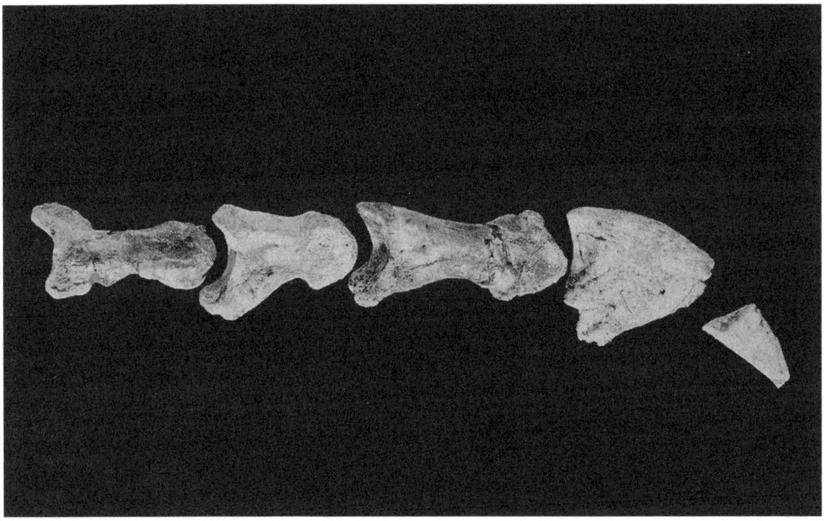

The phalanges of the dinosaur's third finger, recomposed in anatomical connection. © *Natural History Museum of Milan, by courtesy of the Italian Ministry for Cultural Heritage and Activities, Superintendence of Lombardy. Photo by L. Spezia.*

something similar to a whole set of kitchen knives. The paleontologists excavating in the Dinosaur Provincial Park, the largest Canadian dinosaur fossil site, have been known to accidentally injure their hands while extracting *Gorgosaurus* teeth from the sediment: 70 million years after their last bite, these deadly weapons are still capable of causing bloodshed.

The experiments on fresh meat also enabled us to advance the hypothesis that the large carnivorous dinosaurs killed their prey with infected bites. Shreds of meat easily remain trapped in the gaps between the little denticles that form the serrated edges. After remaining a few days in the predator's mouth, these residues would have rotted and possibly infected the wounds of subsequent victims, a hunting strategy used today on the Indonesian island of Komodo by giant monitors, or Komodo dragons.

Tyrannosaurs were not the only carnivorous dinosaurs armed with serrated teeth. Pointed, laterally flattened teeth with sharp edges consisting of small aligned denticles are found in almost all the theropods, from the most ancient *Eoraptor* and *Herrerasaurus* to the ceratosaurs, from the carnosaurs to the coelurosaurs (including *Scipionyx*). Similar serrated edges can also be found on the teeth of other carnivorous amniotes (dimetrodonts, "thecodonts," phytosaurs, mosasaurs, and monitors) and other predatory vertebrates, such as carcharodontid sharks and ziphiid whales. Consequently, this is not a feature that is particularly useful from the systematic and phylogenic standpoint, but it is highly indicative of the phenomenon of adaptive convergence, whereby structures that are particularly advantageous from a functional standpoint are developed independently by different animal groups. The advantage offered by serrated teeth is twofold: they allow the animal not only to grip prey but also to cut its flesh. It also appears that a serrated cutting edge is more resistant to trauma than a smooth edge, which quickly becomes "blunt."

The only tooth found at Saltrio possesses serrated edges, but for the reasons mentioned above, it cannot provide as much information as other anatomical parts. In number and size, the denticles are similar to those on teeth attributed to megalosaurs. However, this observation should be taken with a grain of salt, because a precise characterization of the fossil remains belonging to the genus *Megalosaurus* has not yet been made. At this stage of knowledge, all that can be said is that because of its pointed and slightly backward-curved shape, its lateral compression, and the presence of denticles on the sharp edges, the tooth of the saltriosaur has the characteristics of the tooth of a very large-sized theropod.

Teeth of another kind left their traces on the bones of the saltriosaur

The only tooth brought to light, 4.5 centimeters long, has serrated edges (a short section on the upper right-hand side is preserved). Like the great majority of theropods, the saltriosaur possessed teeth specialized for efficiently cutting the flesh of its prey. *© Natural History Museum of Milan, by courtesy of the Italian Ministry for Cultural Heritage and Activities, Superintendence of Lombardy. Photo by L. Spezia.*

itself. The carcass of the animal was obviously not immediately covered with sediment, as is also apparent from the chaotic position of the bones, not one of which was found in anatomical connection with the others. The carcass was obviously transported by a river or by flood waters and then remained for a few days on a shallow water sandbank, at the mercy of other animals who feasted on it, Jurassic style. Some bones show tooth marks: the right metatarsal II, in particular, bears a semi-circular notch on the front edge of the main portion of the bone, very similar to the "nibbling" caused by the present-day coral-eating fishes, but it could also have been made by the buccal apparatus of a gastropod mollusk. Other circular erosions on the left humerus are more certainly ascribable to mollusks rather than to fishes.

The scientific study of the Saltrio dinosaur will allow us to obtain more accurate data concerning its anatomy and its relationship to already known species. What can already be said with certainty is that due to the simple fact that it is the most ancient large theropod with three-fingered hands and a furcula, this Italian carnivorous dinosaur represents a new genus of animal.

## Dinosaurs, Dry Land, and Marble Palaces

The quarry where the remains of the saltriosaur came to light has a glorious history. It has been employed for almost four hundred years for a far more noble use than its present one. Today quarrying produces gravel for roadbeds, but in the past, Saltrio stone was exploited to obtain precious building stone. In the historic past, what is now simply a rock face disemboweled by mechanical equipment on the southern slopes of Monte Orsa above the village of the little township of Saltrio was an imposing and suggestive place with a sumptuousness all of its own. The old quarry consisted of large cavities with vaulted ceilings supported by natural pillars made of rock that had been left in place.

As early as the seventeenth century, Saltrio stone, like that of nearby Viggiù, was used extensively as building material in the area of Varese, as well as in numerous cities in the north of Italy. For example, it was used to build the upper portions of the façade of La Scala Theater, the Galleria Vittorio Emanuele, and the Royal Palace in Milan; part of the Mole Antonelliana and the train station in Turin; and the Colleoni Chapel in Bergamo. Even earlier use of this stone is indirectly documented in monuments built in the twelfth century and on tombstones dating back to the ancient Romans. The mining activities gave rise to a flourishing local artistic industry that included stone-cutters, marble-workers, and skilled sculptors. The stone from Lombardy was preferred over the travertine stone from Latium because it was much finer grained and lighter in color. Above all, it was more resistant to deterioration by atmospheric agents.

The large cavities and excavation tunnels followed the dip of the strata and opened out at increasing altitudes, ranging from 600 to 750 meters above sea level. The "money streak," free from flint inclusions, was approximately 20 to 30 meters thick. In more recent times, modern excavating equipment has made it possible to reach and extract deeper and overlying levels by opencast mining. Consequently, dynamite and powerful excavating machinery have demolished most of the cavities and pillars and worked out the quarry within the space of just a few years. Mining activities have been reduced since 1996.

According to studies carried out by geologists, the Saltrio Formation belongs to the lower part of the Calcari Selciferi Lombardi unit, which is of Sinemurian-Pliensbachian age (200–190 million years ago). The Saltrio Formation possesses the characteristics of a dolomite limestone with fine loamy laminations. To be more precise, it is "encrinite," a hazel-gray–colored limestone composed primarily of the crushed remains of the plates that formed the external skeletons of echinoderms such as sea urchins, starfishes, and sea lilies. The fossil fauna it contains, however, is much more plentiful (over 100 known species) and has been studied since the last century. In addition to echinoderms, there are ammonites, nautiloids, squids, gastropods, bivalves, brachiopods, bryozoans, shark teeth, and isolated ichthyosaur teeth.

All the ammonites (no fewer than twenty-one species), except for two forms that are also present in the Hettangian (the first stage of the Jurassic Period), belong exclusively to the Sinemurian; *Arnioceras semicostatum,* in particular, is an index fossil typical of the basal part of the Sinemurian and has therefore enabled precise dating of the Saltrio Formation.

From the mineralogy, the sedimentology, and the type of fossils found in the Saltrio Formation, it was clear from the very first studies that these deposits were formed in a marine environment during a warmer climatic regime than the present one, on the bottom of a rather shallow sea not far from dry land. The blocks in which the bones of the saltriosaur were embedded also contained a considerable quantity of glauconite, a greenish-blue mineral considered by geologists to be a bathymetric (sea-depth) indicator. Glauconite currently forms in very shallow waters along the coast during periods of limited silting. The presence of glauconite proves that there must have been dry land in the immediate vicinity. On the other hand, emersion of the region of the province of Varese after the end of the Triassic is proved by traces of erosion on the rocks of that time (Rhaetian stage), by the presence of "red soils" (sediments containing iron and aluminum oxide, which form only in the presence of air), and unconformities (absence of sedimentary rock) referable to the Hettangian stage.

Several fossil sites, which the sedimentologist Alberto Lualdi from the University of Pavia recognized as being coeval to the Saltrio Formation, recently came to light on the eastern shore of Lake Maggiore, near the village of Arolo. The correlation was possible due to the discovery of three species of ammonites of the genus *Arnioceras,* which are considered to be index fossils typical of the Lower Sinemurian stage. Besides the ammonites, these sediments contain plant remains belonging to five different species, four of which are terrestrial: cycadeoids of the genus *Ptilophyllum* (strange plants similar to palms, known only in the fossil state and without living descendants), tall conifers of the genera *Pagiophyllum* and *Brachyphyllum,* and several Cheirolepidiaceae. This is an extinct family of conifers that were widespread during the Lower Jurassic and which, as their small, scaly leaves imply, were adapted to warm, arid climates. The presence of cycadeoids indicates the existence of a "transitional" coastal environment with brackish water, between the dry land and the sea. In fact, during the Lower and Middle Jurassic, the genus *Ptilophyllum* occupied the same ecological niche as present-day mangroves.

In short, paleobotanical studies have proved that about 200 million years ago, in a geographic area corresponding to northwest Lombardy, there were dry lands overlooking a sea with irregular shallow coasts. Most of the territory in the region of Varese and Lugano formed a "structural high," an isthmus of unknown extension that was delimited to the west and the east by two stretches of water, called respectively Monte Nudo Basin and Monte Generoso Basin. To the southwest, this sea opened out into a deep ocean, known to geologists as the Ligurian-Piedmontese Ocean, which in actual fact was a gulf of the vast Tethys Ocean.

Until a few years ago, the most common image of this ancient environment, obtained exclusively from geological studies, was of a predominantly marine environment, dotted here and there with small low-lying islands, similar to present-day tropical atolls. The areas of dry land were consequently considered to be very limited. Plants, even very large ones such as araucarias, can survive in limited spaces, taking root in favorable soil from which they do not need to move. But a tiny island is definitely not enough for an 8-meter-long carnivorous dinosaur. Since the large theropods were at the top of a complex food pyramid composed of an imposing biomass, the presence of a large terrestrial carnivore necessarily implies the existence of a whole series of living creatures, from the producers of plant materials to their primary and secondary consumers, from the forests to the large herbivorous dinosaurs that constituted the food necessary to enable animals such as the saltriosaur to survive. The first Lombard dinosaur must not, and indeed could not, have been alone. Where the others were, it is still too soon to say, or perhaps we will never know. The remains of the saltriosaur definitely were transported, but we should not necessarily presume that they came from far away. In the same Saltrio Formation, there are fossil traces of ancient beaches (as we mentioned previously, the levels containing polygonal ground and surf ripple mark a precise shoreline): there could be traces of the passage of another saltriosaur and of other types of dinosaurs on these tidal flats.

At the present stage of knowledge, it is difficult to know whether 200 million years ago the areas of dry land in northern Lombardy were in some way linked to the contemporary coastal region of Rovereto. However, it is highly likely that relatives of the saltriosaur roamed there too. Consequently, there may be more than just coelophysoids among the numerous footprints of large carnivorous dinosaurs at Lavini di Marco. And among the prey of the Lombard dinosaur, there could quite easily be sauropods and ornithopods similar to those that roamed the Trento Platform.

# 6

# THE BESANOSAUR AND MARINE REPTILES

THEY INHABITED THE oceans more than 230 million years ago, even before the dinosaurs began their march over dry land to conquer the world. The ichthyosaurs swam swiftly and sinuously through the water, their shiny streamlined bodies recalling those of the shark and the dolphin, but they were neither fishes nor cetaceans: they descended from primitive terrestrial reptiles that had adapted perfectly to living in water, transforming their legs into fins and acquiring a hydrodynamic shape. One of the most extraordinary and well-preserved ichthyosaurs came to light in Italy in the province of Varese in 1993. It is called the besanosaur. Its 6-meter-long fossil skeleton emerged from an 11-square-meter slab of rock, still articulated, offering a unique view of the largest and most complete marine reptile fossil ever discovered in Italy.

While dinosaurs are a novelty for Italy, other reptiles that lived in the region during the same era have been known for decades, some even for a couple of centuries. In fact, their discovery resulted in some of the earliest publications in the history of paleontology. Ichthyosaurs and many other aquatic reptiles found a particularly favorable environment for their evolutionary diversification in the vast Tethys Sea, which obstructed and isolated the dinosaurs. They left numerous fossils in the widespread marine deposits that outcrop throughout the Italian peninsula. The age of these deposits ranges from the Middle Triassic, about 240 million years ago, to the Upper Cretaceous, which ended 65 million years ago. However, not all the localities have been studied in depth or have yielded equally abundant and significant fossils. Skeletal remains of the spectacular aquatic monitors of the Cretaceous known as mosasaurs, veritable sea monsters sometimes exceeding 10 meters in length, came to light in Italy in the Venetian and Emilian Apennine Scaglia Formation. Remains of ichthyosaurs have been found in the Dolomites, in Carnia, and in the Apennine region of the Marches, but in terms of abundance and state of preservation, the best-known are those of Besano.

## Digging into the Triassic

The rocks that outcrop north of Varese near the village of Besano on the slopes of Monte Pravello and Monte Orsa and in the adjacent area of Monte San Giorgio in Swiss territory have been known to the local inhabitants for centuries for their capacity to yield an oil that burns easily and an unguent called Saurol (known elsewhere as Ichthyol), which until a few years ago was used as a drug for rheumatism and skin inflammation. Due to the high percentage of organic matter they contain and their blackish color, these stratified rocks have been known since the nineteenth century by the name of Scisti Bituminosi (bituminous shales). Around 1830 there were plans to utilize the shales to extract a gas that could be used as a fuel for illuminating the city of Milan. The project came to nothing, but it served to draw the attention of scientists to the fossils contained in these strange rocks.

The first publication dates to 1847 and was written by the Milanese geologist Giulio Curioni. In 1854, another Milanese geologist, Emilio Cornalia, was the first to describe and depict a Besano fossil, a small reptile that he named *Pachypleura edwardsii,* which came from a geological formation just slightly above the Scisti Bituminosi. The interest aroused by this and other discoveries was so great that in 1863 the Abbot Antonio Stoppani, then president of the Italian Society of Natural Sciences, promoted an excavation aimed at finding fossils in those rocks. This work, which cost all of 500 lire (a quarter of a euro, a figure that nowadays brings a smile to the lips but at that time represented a small fortune), yielded results that fired the Milanese paleontologists with such enthusiasm that fifteen years later another excavation was carried out, financed by the Natural History Museum of Milan and directed by Emilio Cornalia. Since then, the Lombard fossil site, initially believed to be of Jurassic age but later dated to the Middle Triassic (about 235 million years ago), has become known throughout the world for its abundance of invertebrates, fishes, and especially marine and terrestrial reptiles (no fewer than twenty-four species of the latter are known to date).

In August 1943, at the height of the Second World War, the heart

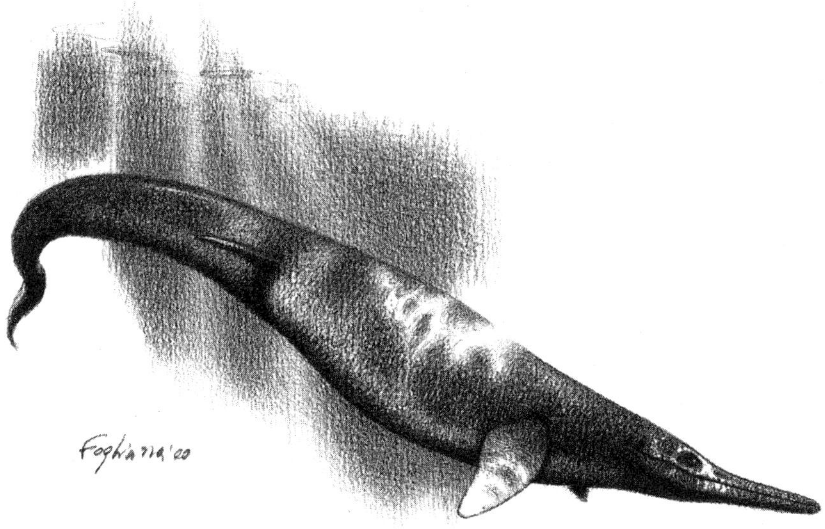


The besanosaur was a primitive ichthyosaur. This can be seen by the lack of a dorsal fin and crescent-shaped caudal fin, which, in contrast, are present in the ichthyosaurs of the Jurassic. It swam easily through the water with an eel-like motion. The limited size of the teeth and the tapered shape of the rostrum indicate that the besanosaur fed on cephalopod mollusks. *Drawing by F. Fogliazza.*

of Milan suffered a devastating air raid. The Natural History Museum was hit by a cluster of incendiary bombs. Hundredweights of bituminous shale burned, due to its high mineral-oil content, together with the museum's furnishings and wooden showcases. The famous collection of Triassic fossils of Besano, the fruits of paleontological excavations and discoveries made during the exploitation of the mine, literally went up in smoke. Only a few specimens of reptiles, which were on loan in Zurich for study purposes, were saved. In 1924, the Zurich Polytechnic had begun a series of studies on Swiss territory in the Scisti Bituminosi and in several levels of the overlying Meride Limestone. Numerous excavation campaigns had yielded an enormous quantity of fossils and had made possible the production of a substantial series of scientific publications.

At the end of the war, the Milan Natural History Museum rose from its ruins. In 1975, paleontologist Giovanni Pinna, following the tradition of the Institute, decided to rebuild the lost collection of Besano. The new excavations, opened in the locality of Rio Ponticelli with the co-

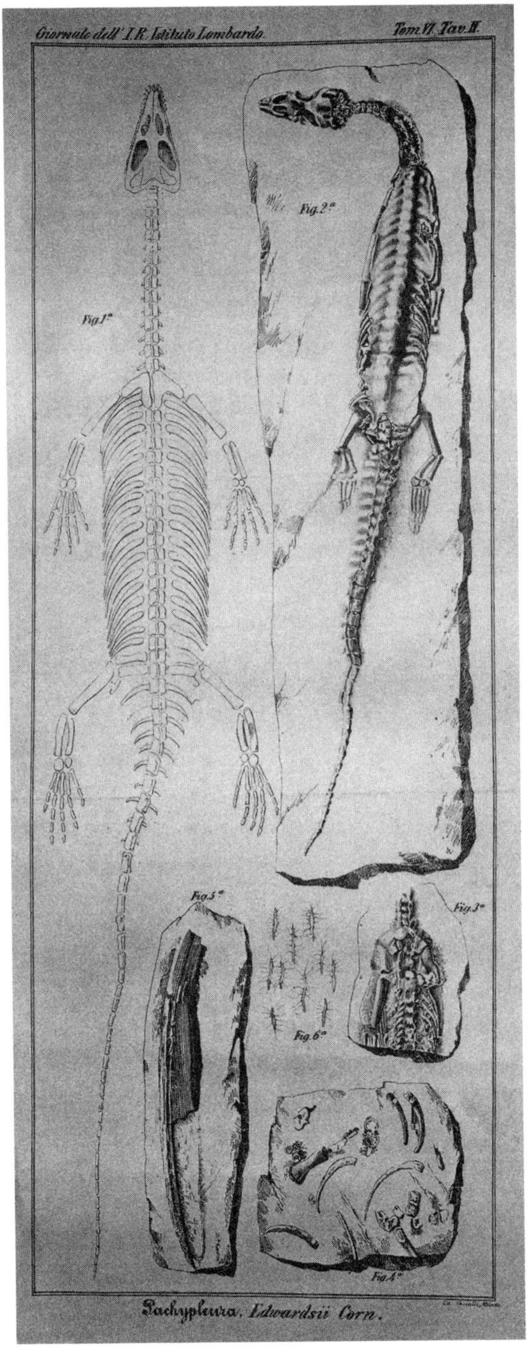


Plate from E. Cornalia (1854) accompanying the description of the first fossil reptile discovered at Besano. It consists of an eosauropterygian, at that time christened *Pachypleura edwardsii* but now attributed to the genus *Neusticosaurus*. *Photo by L. Spezia.*

operation of a team of volunteers, yielded remains of large fishes of the genera *Colobodus* and *Birgeria* as well as a number of small ichthyosaurs of the genus *Mixosaurus*. But the miraculous haul occurred when the same excavation team, still operating under the direction of Giorgio Teruzzi, cast its nets into the new Sasso Caldo site.

Opened in the fall of 1985, the Sasso Caldo (Warm Stone) quarry takes its name from an outcrop of dolomite rocks that emerge from the vegetation into a sun-drenched spot on the mountain. Despite dense undergrowth, volunteers of the Paleontological Group of Besano—especially Claudio del Prato, who was very well acquainted with the neighboring outcrops—"scented" the good conditions of exposure of the shales in that spot. The evenness of the slope and the absence of large faults suggested a certain continuity of the underlying rock layers, and a nearby path offered better access to the site than elsewhere. So, at the rate of about 1 meter a year, the workers began their descent into space and time, systematically removing alternating strata of dolomite rock and bitumen that cropped out one under the other in regular succession over a thickness of a dozen meters and a time span of 13 million years. From the uppermost levels, formed about 227 million years ago during the Lower Ladinian stage, they worked down to the basal levels, dating back almost 240 million years ago to the Upper Anisian. This was the start of the longest excavation campaign ever undertaken, which was to produce the richest collection of specimens that the Milanese museum had ever recovered from a single fossil site. Included was a spectacular ichthyosaur destined to arouse a great deal of comment.

## Hunting for Ichthyosaurs

It was spring 1993. The first warm rays of sunlight of the season did not reach the Sasso Caldo outcrop until midday. It was still bitterly cold on the northern slope of Monte Pravello at an altitude of 800 meters. So the members of the Paleontological Group lit a bonfire. A few weeks earlier, they had been kindled by great euphoria: in the stratum marked number 64, and only a few centimeters apart, lay two complete mixosaurs (small ichthyosaurs) measuring about 1 meter each. To be able to extract them, the workers had to isolate an area measuring 2 by 2 meters, and lifting a 20-centimeter-thick slab of dolomite rock of that size was by no means child's play. They decided to delimit a perimeter around the two ichthyosaurs, remove all the rest of the stratum of dolomite rock, and then slide the isolated slab over the underlying stratum. But stratum 64 was so firmly attached to the fine bituminous level on which it rested that on the left edge of the excavation area, a slice of level number 65 came away together with the thick slab of dolomite rock. The field crew realized immediately that the fracture had not occurred on a random plane. Embedded in the black bitumen, there gleamed a dark brown spindle shape divided along its midline by a zigzag line. On closer inspection, the zigzags turned out to be teeth regularly spaced one centimeter apart. The spindle was the front end of the mouth of an enormous ichthyosaur.

The marine reptile was practically trapped beneath the two mixosaurs, which consequently resulted in many weeks of additional work. Since the dolomite rock strata alternated with shale strata, it was necessary to excavate first with heavy sledgehammers, divaricating wedges, and a motor-driven demolition hammer into the hard, thick dolomite rock and then, after a sufficient area of bituminous shale had been freed, to very carefully remove these slabs with chisels and spatulas.

The excavation team, coordinated by paleontologists from the Natural History Museum of Milan, took the whole summer to complete the recovery of the huge ichthyosaur, or rather, the stratum in which it was enclosed. In fact, not knowing how big the specimen was or whether it was complete, they decided to take the entire layer. Initially they attempted to detach the shale from the underlying dolomite rock by inserting chisels between the two levels, but this caused too many natural fractures in the finer slabs, which, in addition to shattering some of the bones, would have produced fragments that could no longer be reassembled. In order to extract an area as wide as possible without breaking it up into a thousand pieces, they preferred to make use of a portable shear, a motor-powered diamond saw, with which rectangular crosscuts were made. Immediately after they were removed, the slabs were marked with symbols so that their position within the huge "mosaic" could be remembered. To avoid damaging the fossil, for months and months the team had to make do with observing its bones only in cross-section, along the shear plane of the stratum that enclosed the ichthyosaur as in a jewelry box. Despite this meticulous work, the slabs often split up still further, so all the pieces had to be patiently reassembled and immediately packed.

Ichthyosaurs are by far the reptiles most frequently found at the Besano fossil site. The name, which means "fish-lizard," describes the characteristics of these animals perfectly. Although ichthyosaurs adapted to living in water by transforming their legs into four flippers, their internal anatomy and physiology remained that of terrestrial reptiles and not that of fishes. (For example, they breathed with lungs and had to rise to the surface often so as not to drown.) The skull, which tapered into a beak-like process called a rostrum, had enormous round orbital cavities, each one housing a bony ring, which is actually the ossified sclera that reinforced the margin of the eye.

Thanks to traces of soft tissue in several ichthyosaurs found at the Jurassic fossil site of Holzmaden, in Germany, exceptionally well preserved in the form of a carbon film, it is possible to reconstruct the outline of the body of these animals. In addition to the four flippers, which were supported by bones of the limbs appropriately reduced in length and strengthened by an increased number of phalanges (there are species that have up to 100 phalanges per flipper), ichthyosaurs were also provided with a cartilaginous dorsal fin and a crescent-shaped caudal fin. The lower lobe of the latter was reinforced by the downward-bending last section of the spinal column, while the upper lobe was composed entirely of cartilaginous tissue.

Ichthyosaurs occupied the ecological niche that today belongs to dolphins, populating the surface waters of the seas and oceans of the Mesozoic Era until the Cretaceous Period. They died out before the dinosaurs, approximately 80 million years ago. The most widely distributed forms during the Triassic Period are grouped into two families, the mixosaurids and the shastasaurids, both well represented in the Scisti Bituminosi Formation. In fact, mixosaurids are typical of the outcrop: these ichthyosaurs were first discovered at Besano and are particularly abundant there. For the most part the skeletons, usually complete but rarely articulated, belong to the genus *Mixosaurus.*

*Mixosaurus cornalianus,* so called by Emilio Repossi in 1902 in honor of Emilio Cornalia, rarely exceeded 1 meter in length and was the most common species. The mixosaur, once considered the most primitive ichthyosaur, in fact possessed an anatomical structure highly adapted to water in comparison to its predecessors, such as *Utatsusaurus,* which dates back to the Lower Triassic of Japan. However, compared to more derived forms, which appeared during the Jurassic and Cretaceous, the mixosaur had a tail well developed only on the lower lobe and probably lacked a

Besano (Varese), spring 1993. Sasso Caldo quarry. Removal of the solid dolomite rock making up stratum number 64 (*top*) to free the underlying bituminous level, which was only 3 centimeters thick (*bottom*), containing an enormous ichthyosaur. To avoid damaging it, the entire stratum number 65 was cut and recovered.
*© Natural History Museum of Milan, by courtesy of the Italian Ministry for Cultural Heritage and Activities, Superintendence of Lombardy. Photo by C. Dal Sasso.*

dorsal fin, while inside the paired fins it still retained a five-fingered limb. As shown by recent discoveries of larger individuals (up to 180 centimeters in length), apparently there were other species of mixosaurs in Besano, but these specimens have not yet been studied.

The biggest and most spectacular ichthyosaur that has ever been found in the Scisti Bituminosi is that specimen which, after being removed from the Sasso Caldo quarry enclosed in no fewer than thirty-eight slabs, did not reveal its identity until its arrival in the heart of Milan, and then only after a very singular investigation.

## An X-Rayed Fossil

Thanks to the cooperation of Ospedale Maggiore, the major city hospital of Milan, we were able to X-ray all the slabs taken from stratum number 65. X-rays are highly penetrating and allowed us to see through the rock, almost as though it were transparent, due to differences in density between the fossil bones and the sediments encasing them. Under the guidance of Professor Giorgio Cosmacini, then head of the Radiology Department of the Milanese Polyclinic, together with Chief Technician Gianni Tafuni, we slowly viewed the radioscopic images on a monitor and then fixed them onto radiographic plates. The enigma was solved, radiograph after radiograph. Like a jigsaw puzzle, one image had to fit with another. The huge reptile gradually began to take shape. The head was long and pointed and the thorax was wide, with the ribs maintaining the position they were in when the animal was alive. The spinal column, composed of no fewer than 201 disks resting one on the other like the pieces of a game of Chinese dominoes, seemed to be never-ending. At this point, we realized the value of the discovery. The bones of the ichthyosaur were almost all in anatomical connection and extended over such a wide area that it took 145 radiograms, each 35 by 45 centimeters, to map out the whole skeleton. The images were redrawn and assembled on a sheet of transparent plastic, finally revealing the shape and size of the animal. The result was an 11-square-meter area constituting the complete skeleton of the new reptile, which from head to tail measured 580 centimeters. Also revealed was a small ichthyosaur of the genus *Mixosaurus,* which was about 70 centimeters long.

But the most thrilling moment of the X-ray examination was when the thorax of the huge ichthyosaur appeared on the monitor. The radiologist was absolutely astonished. It looked as though his patient had

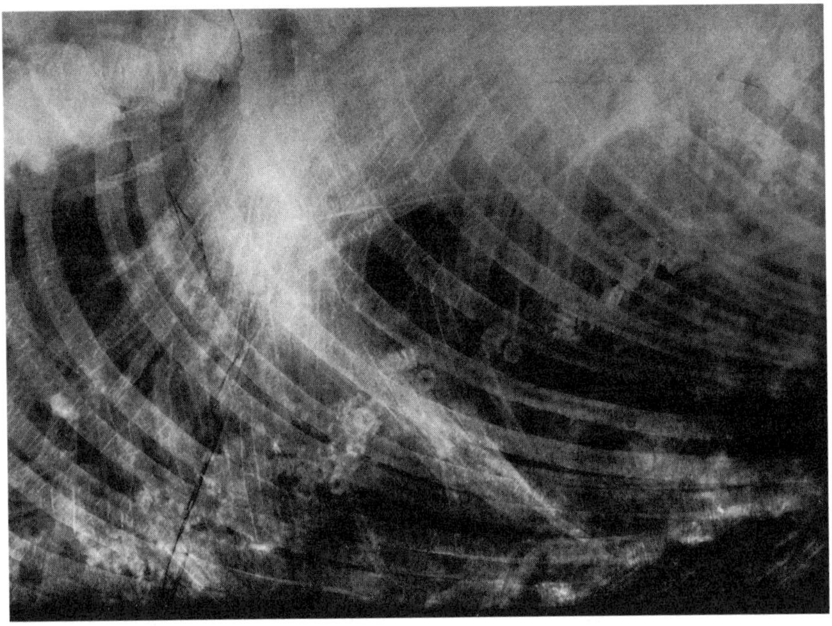

By means of 145 X-rays, taken by the Ospedale Maggiore of Milan, it was possible to see the big ichthyosaur in transparency through the rock. Here we see a detail of the thorax; the tiny disks are the vertebrae of the embryos. *© Natural History Museum of Milan, by courtesy of the Italian Ministry for Cultural Heritage and Activities, Superintendence of Lombardy. X-ray by G. Tafuni. Photo by L. Spezia.*

swallowed dozens of tiny coins. In actual fact, it was evidence of a drama that had taken place 235 million years ago. The "little coins" contained inside the thoracic cavity were the vertebrae of smaller ichthyosaurs, a discovery within a discovery. They were almost certainly embryos—new, unfinished lives that this almost 240-million-year-old marine mother had been unable to deliver in her own time. Instead they had been delivered to the future world through her death and fossilization.[1]

Ichthyosaurs were so well adapted to marine life that they did not need to come out of the water even to reproduce. The fact that these reptiles were viviparous—they did not lay eggs but gave birth to fully developed offspring in the open sea—was known only because of a few exceptional discoveries previously made at the fossil site of Holzmaden, near Stuttgart. (However, these were of a more recent period, Lower Jurassic, approximately 175 million years ago.) From the middle of the nineteenth century to the present, the Holzmaden site has yielded about three thousand ichthyosaurs. According to an estimate made by the Canadian paleontologist Christopher McGowan, out of these three thousand there were only eleven pregnant females.

After X-ray examination, the big ichthyosaur of Besano ended up on the operating table in the Laboratory of Paleontology of the Natural History Museum of Milan, where it underwent a long and delicate operation of reassembly and preparation. After being extracted, and now no longer subjected to the pressure exerted by the overlying rock strata, the bituminous shale tended to dry out and buckle like panels of wet wood. Unless prompt action had been taken to consolidate them, the fractured slabs would have become more difficult to reassemble. With the help of binocular microscopes and very fine-pointed steel instruments, three specialized technicians, working painstakingly by hand, millimeter by millimeter, began to remove the rock matrix in which the huge fossil was embedded. Since it was not possible to extract the skeleton completely due to the strong compression of the sediment in which it was sandwiched, the bones had to be exposed from one side only. The preparation lasted five years. Twenty-one thousand square centimeters of bituminous rock were worked under the microscope over a period of 16,500 hours. In the meantime, the specimen was studied by means of the X-ray images. In September 1999, it was presented to the press and exhibited to the public at the Museum of Natural History of Milan and the Fossil Museum of Besano.

## Face to Face with the Besanosaur

It is significant that after more than 150 years of exploration, a fossil site thought to be known to all and sundry should yield a new species of reptile represented by a 6-meter-long animal. In paleontology there is still much to be discovered, perhaps more than we can imagine, and nothing can ever be taken for granted.

Apart from this lesson, there are several reasons why the discovery of the big ichthyosaur of Besano is of great importance. First and foremost, this specimen is unique for science because, even though thousands of ichthyosaurs are known, it represents a new genus and a new species, of which no other specimen has yet come to light anywhere else in the world. The paleontologists who studied it christened it "besanosaur" in honor of the locality where it was found. Due to the unusually long and slender rostrum, the species was given the name *Besanosaurus leptorhynchus* (literally meaning "thin-billed reptile of Besano").[2]

Second, it is the largest ichthyosaur ever found in Italy, and one of the oldest and best-preserved in the world. Indeed, as far as ichthyosaurs from the Middle Triassic are concerned, no other skeletons as big and at the same time as complete have ever come to light in any other locality. The largest known specimen, belonging to the genus *Cymbospondylus,* is 10 meters long and is in Nevada, but it is missing part of the tail and the limbs. In Europe, a single specimen of *Cymbospondylus* was found in 1927 in the Scisti Bituminosi, on the Swiss side of the Alps, and is now kept in Zurich. It is an articulated skeleton 230 centimeters long but preserved only in the front half. The genus *Cymbospondylus* was widespread in the seas throughout the world, but the Swiss specimen belonged to a particular species called *Cymbospondylus buchseri.* This marine reptile was characterized by a relatively small skull, an elongated body, limbs transformed into large, narrow flippers, and a long and almost rectilinear tail that was devoid of a falciform fin. Rather than a fast swimmer, it must have been a predator that lay in wait and pounced upon its prey. *Cymbospondylus,* until quite recently considered to be related to the genus *Shastasaurus* and included in that family, is now considered to be a closer relative of the small mixosaurs.

Anatomical study of the spectacular besanosaur revealed that it belongs to the shastasaurids. Shastasaurids are primitive ichthyosaurs, typical of the Triassic period but not very well known due to the incompleteness of their fossil remains, which are found mainly in California and Germany. The scientific importance of the besanosaur lies in its outstanding completeness. Not even a single vertebra of its spinal column is missing. In addition, the presence of the embryos offers the possibility of studying the ontogeny of ichthyosaurs (i.e., how they changed during development). The besanosaur also possesses characteristics halfway between the genera *Shastasaurus* and *Cymbospondylus.*

It is probable that, like other primitive ichthyosaurs, *Besanosaurus leptorhynchus* had still not developed the dorsal fin and the falciform caudal fin (anatomical characteristics that, in contrast, are present in the ichthyosaurs of subsequent periods). It can be inferred from the shape of the body that the besanosaur also swam with an eel-like motion, swaying from side to side as fishes do, and not dorsoventrally, as dolphins and whales do. Rather than being designed for high speeds and long dis-

The fossil skeleton of the besanosaur is 5 meters and 80 centimeters in length. Here it appears in bas-relief on the 11-square-meter slab of bituminous shale in which it was compressed, after its preparation and reassembling, which took as long as 16,500 hours. *© Natural History Museum of Milan, by courtesy of the Italian Ministry for Cultural Heritage and Activities, Superintendence of Lombardy. Photo by L. Spezia.*

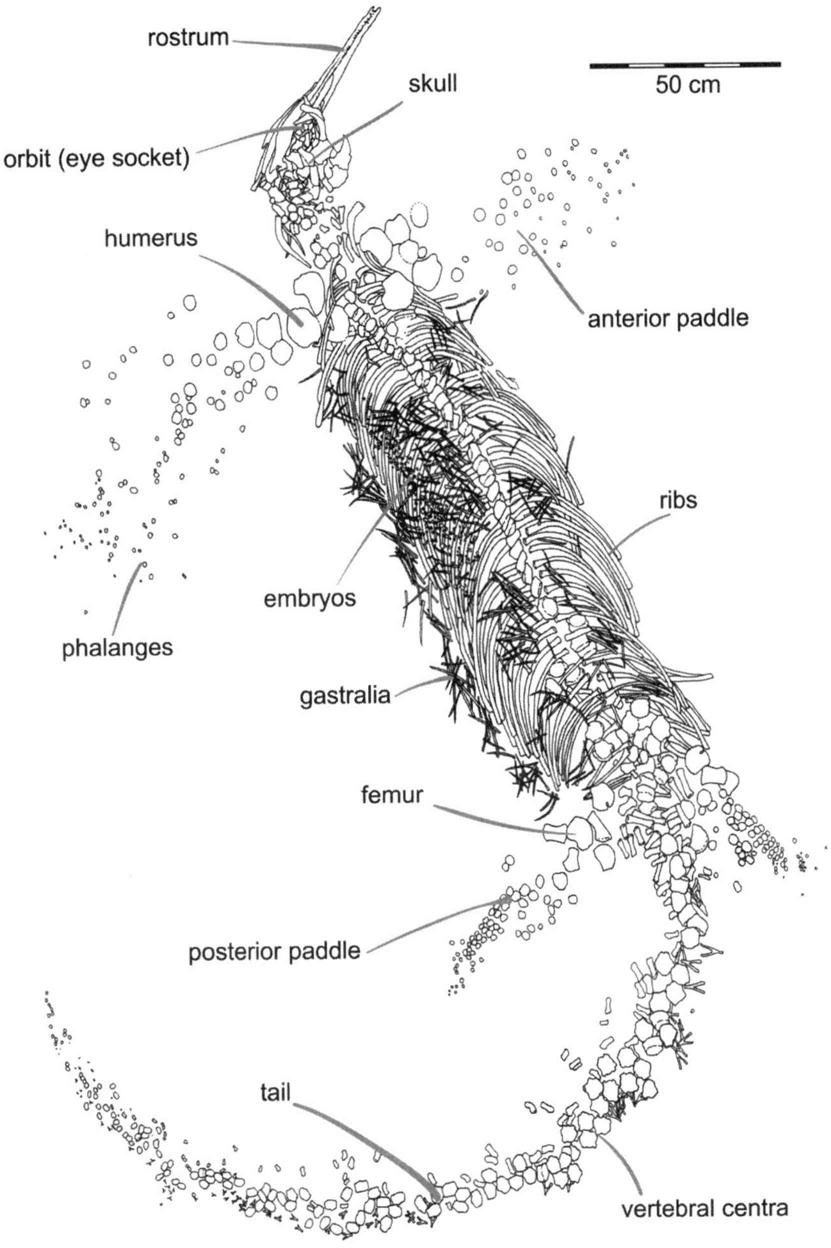


Almost all the bones of the besanosaur are in anatomical position. The phalanges of the anterior flippers and the vertebrae of the tail are spaced slightly apart because shortly after the carcass was deposited on the seabed, the cartilages that held them together underwent decomposition. *From Dal Sasso and Pinna 1996, modified. Graphics by F. Nodo.*

tances, its body was built for short, sudden lunges and possessed good maneuvering capabilities. Its small teeth and tapered rostrum indicate that the besanosaur fed almost exclusively on squid.

Triassic ichthyosaurs showed an evolutionary trend toward loss of the terrestrial adaptations of their ancestors in favor of greater specialization for swimming. The besanosaur still had a nearly rectilinear tail, very like that of a lizard, and an elongated thorax. The large number of presacral vertebrae is a primitive character also shared by *Cymbospondylus*. These enabled them to perform much more sinuous movements than the Jurassic and Cretaceous ichthyosaurs, which, instead, had a compact shape and swam in leaps, like tunas, and were therefore faster swimmers made for life in the open sea and for covering great distanccs.

Recently, the Japanese paleontologist Ryosuke Motani, a world expert on ichthyosaurs, recognized in the anatomy of the besanosaur many of the characteristics that one would expect to find in the primitive shastasaurs. He surmised that the specimen discovered in 1993 represents a new, very ancestral subfamily that he called the Besanosaurinae. The besanosaurines, which lived during the Middle Triassic, must have given rise to the shastasaurines, a more highly evolved subfamily that includes larger forms of ichthyosaurs that spread throughout the seas of the world during the Upper Triassic. Among the latter, *Shonisaurus* was a genuine giant. Discovered in Nevada, it was no less than 15 meters in length and weighed up to 30 tons. Its skull alone measured 3 meters. This animal, much bigger than the dinosaurs that inhabited the supercontinent Pangea during that period, probably had a lifestyle similar to that of the present-day large toothed whales, such as the killer whale. *Shonisaurus* was the biggest ichthyosaur of all time. However, contrary to our besanosaur, the shonisaur was reconstructed on the basis of numerous fragmentary skeletons.[3]

Whatever the precise relationship between the besanosaur and its fellow creatures of the Middle Triassic, the existence of a complete skeleton dating back to 235 million years ago offers the possibility of opening a unique window onto the evolution of this group of marine reptiles.

## A Tropical Sea

The besanosaur lived in an environment totally different from the one we see today in the same area. During the Middle Triassic, the Alpine region was covered by a vast expanse of warm sea, with a flat seabed composed of sediment derived from disintegrated shells and calcareous skeletons of algae, corals, and mollusks. It was a carbonate platform environment, similar to that of Pietraroia but geologically more active, a shallow sea scattered with small islets and erupting volcanoes, which, on emerging at the surface, gave rise to chains of volcanic islands. In the middle of this sea, in a small area corresponding to the surroundings of Besano and Monte San Giorgio, there was a basin with a maximum diameter of 10 kilometers and a depth of about 100 meters. To the north, toward the area now occupied by Monte San Salvatore, the basin was bordered by a reef composed of calcareous algae, beyond which, toward Lugano, lay a lagoon from which volcanoes emerged.

The Besano Basin was a sort of huge, deep pool surrounded by semi-emergent areas and shallow depths that restricted the circulation of currents, especially close to the seabed. Organic matter gradually accumulated and decomposed there, contributing to a lack of oxygen already caused by the limited exchange of water. The marine organisms lived in the shallowest reaches of the basin, which by contrast were regularly oxygenated. Since the seabed was devoid of life, the remains of plants and animals that were deposited there were not destroyed by other organisms; nor were they dismembered and scattered by the current. This explains the excellent state of preservation of the fossils of Besano. In the outcrop, the strata richest in vertebrate remains are the bituminous layers. They were deposited over long periods, during which—since currents laden with sediment could not reach the basin—only organic matter, composed mainly of a large quantity of microscopic algae, slowly sank to the bottom. Fossil vertebrates are concentrated in the shale, because during the long period required to form a bituminous layer, a great number of carcasses arrived on the bottom. However, every so often natural events occurred, such as storms or hurricanes, which removed calcareous sediment from the shallow edges of the carbonate platform. Part of this sediment was discharged into the basin and quickly deposited on the bottom by means of turbidity currents, thus forming the dolomite rock layers, which are thicker and contain fewer fossils. The Besano Formation is the result of an alternation of prolonged periods, characterized by very little sedimentation on the bottom and the formation of the bituminous levels, and shorter periods, during which sediment from the carbonate platforms was deposited in the basin. On average, in the stratigraphic series of Besano, 1 meter of rock corresponds to about 500,000 years.

The outcrop contains not just the remains of marine organisms but also evidence of what life must have inhabited the islands and dry land surrounding the basin: a variety of organisms reflecting almost entire ecosystems of more than 230 million years ago, from plants to animals, from marine invertebrates to terrestrial vertebrates, and from fishes to reptiles. The quantity of discoveries is also remarkable. From the nineteenth century to the present, the site has yielded thousands of fossils, which permit us to reconstruct the life forms that inhabited the semi-submerged Italy of the Middle Triassic Period.

The fossil remains of terrestrial plants show that the vegetation on the dry lands close to the ancient sea of Besano consisted primarily of primitive conifers similar to the present-day araucaria or monkey puzzle tree (genus *Voltzia,* typical of the Triassic). A fossil found a few years ago indicates that there were scorpions among the terrestrial invertebrates, and a great variety of insects must have flourished in these subtropical forests.

Marine mollusks, which are particularly abundant in the dolomite rock levels, indicate that the well-oxygenated shallow depths surrounding the Besano Basin teemed with invertebrates, especially bivalves and cephalopods. The most numerous of the latter were the ammonites, which were widespread during the Triassic after narrowly avoiding extinction just a few million years previously, toward the end of the Paleozoic Era. By studying the stratigraphic distribution of the different species of ammonites, it has been possible to date the site to the exact boundary between the Anisian and Ladinian stages and to determine its correlation with other fossil-bearing outcrops in the Alpine area, especially in eastern Lombardy and the Dolomites. Thanks to radiometric dating of the fine layers of volcanic ash that intercalate with the bituminous shale, it was possible to establish that the absolute age of the Besano Formation is 232 million years, with a margin of error of 5 million years either way.

The most abundant vertebrates are the fishes. They are preserved at a diversity rivaling that of modern reef environments. There are cartilaginous fishes (or Chondrichthyes, the class that includes the present-day sharks) as well as bony fishes (or Osteichthyes, the class to which most living fishes belong). The sharks of the Sea of Besano were very distinctive: they belonged to an extinct group, the hybodonts, and had stubby, triangular, or hemispherical teeth, unsuitable for cutting but useful for grinding mollusk shells and the rigid exoskeletons of crustaceans. Studying the bony fishes of Besano is fundamental for understanding the evolution of this class of vertebrates. In fact, the Middle Triassic is characterized by the presence of a large number of "experimental forms," some of which bear witness to the transition from a primitive type of fauna, which had characterized previous periods, to another, more advanced type. The little perleidiforms and peltopleuriforms, for example, show innovative anatomical characters that suggest an improvement in their ability to swim and capture food.

## A Unique Window onto the Past of Reptiles

While fishes are the most abundant fossils, the best-known forms are undoubtedly reptiles. Besides ichthyosaurs, the Besano Formation and the overlying Meride Limestone yield such an enormous variety of other reptiles that it is possible to reconstruct an entire ecosystem, a snapshot more vivid and complete than ever before of how the seas and lands of Lombardy must have been populated in an age so far removed from ours, some 230 million years ago.

Shoals of ichthyosaurs swam in the open sea, while close to the shore, waters teemed with semi-aquatic reptiles such as pachypleurosaurs, placodonts, and thalattosaurs. Archosaurs and other animals, such as protorosaurs, roamed along the shores and in the hinterland. All these groups are extinct. Many of them are limited to the Triassic and bear witness to a biodiversity that the reptilian class has now lost—a variety of evolutionary experiments that marks the start of the Mesozoic Era, with good reason nicknamed the Golden Age of Reptiles.

Despite the fact that the Besano fossil reptiles have been known since the middle of the nineteenth century, explorations currently underway are still leading to the discovery of specimens belonging to new species. These discoveries further enhance the already exceptional diversity offered by this site, which is arguably the most important worldwide for vertebrate faunas from the Middle Triassic. However, at Besano there are no dinosaurs, even though the oldest known dinosaurs—found only in Argentina, Brazil, and Madagascar—are more or less the same age as the fossils from the Lombard site.

Close to shore, the Sea of Besano abounded in eosauropterygians, semi-aquatic reptiles typical of the Triassic Period. Medium to small in size and shaped like lizards with an elongated neck and webbed feet, eosauropterygians are considered to be the ancestors of plesiosaurs, marine reptiles with the stocky body and snakelike neck that legend attributes to the Loch Ness monster. The most interesting feature shared in common by these animals is highly developed bones of the pectoral and pelvic girdles, which served as attachments for muscles that moved the legs. In some species, the ribs were enlarged to weigh down the body of the animal and overcome its tendency to float.

In the Besano–Monte San Giorgio area, there are two kinds of eosauropterygians: the pachypleurosaurs and the nothosaurs. The former include the serpianosaur (*Serpianosaurus mirigiolensis*) and a species of reptile described by Emilio Cornalia in 1854 as *Pachypleura edwardsii,* now known by the name of *Neusticosaurus edwardsi.* The pachypleurosaurs' middle ear, shaped similarly to that of terrestrial reptiles, implies that they were not able to descend to very great depths. They fed on small invertebrates or small fishes and were in turn prey for the large lariosaurs, as is proved by the fossils of fecal remains of large reptiles containing their bones.

The family of pachypleurosaurids is the best known scientifically. By studying more than four hundred fossil skeletons housed in the Museum of Zurich, German paleontologist Martin Sander was able to make an in-depth study both of sexual dimorphism and of the life cycle of the neusticosaurs. By counting the growth rings made visible by sectioning the bones, he established that the neusticosaurs lived from six to nine years. At birth they measured 5 to 6 centimeters, and as adults they reached a length of 120 centimeters. From birth to four years of life, they were all identical in body form; then the proportions of the forelimbs changed according to whether they were males or females.

Particularly different, both in shape and in size, was the ceresiosaur, a nothosaur belonging to the family of the lariosaurids. In fact, in body structure, the ceresiosaur is slightly more primitive than the plesiosaurs, having a very long neck and extremely sturdy paddle-shaped limbs, and is undoubtedly the nothosaur best adapted to living in water. The ceresiosaur was perhaps one of the predators of the little neusticosaurs: a fine specimen measuring 230 centimeters, discovered on Monte San Giorgio in 1933, lies on a slab together with various neusticosaur skeletons and a coprolite containing neusticosaur bones. *Ceresiosaurus* is now considered a synonym of *Lariosaurus,* a genus described for the first time in 1847 on the basis of fossil remains discovered in Perledo, near Varenna (Lecco).

A few years ago, Olivier Rieppel of the Field Museum of Chicago and Rupert Wild of the Museum of Stuttgart proposed a synonymy for the big nothosaur *Paranothosaurus amsleri,* an almost 4-meter-long skeleton of which was found in the Scisti Bituminosi of Monte San Giorgio in 1932. According to these paleontologists, the huge skeleton can be attributed to *Nothosaurus giganteus,* a species described in 1834 in Germany. In fact, the genus *Nothosaurus* is found in profusion in the rocks of the German Middle Triassic, which would appear to indicate that there may have been faunal exchanges between the Alpine area and Central Europe during that period.

The placodonts, which were related to the eosauropterygians but were fully adapted to life in the open sea, also lived exclusively during the Middle Triassic. They had a massive body, with a wide, dorso-ventrally compressed head, enormous flat teeth capable of crushing the shells of marine invertebrates, and short, stubby limbs. Nearly all were provided with a protective shell similar to the carapace of a turtle. The remains of two very distinct genera, *Cyamodus* and *Paraplacodus,* are found in the Besano Formation.

The askeptosaur (*Askeptosaurus italicus*), which belongs to another extinct order, the thalattosaurs, was a large semi-aquatic predator about 2 meters long. It was very similar to a crocodile but more specialized for life in the open sea, with posteriorly located nostrils on top of the snout to facilitate breathing while nearly submerged, short limbs, and a laterally compressed tail for swimming. It fed on fish and probably also small reptiles. It was described for the first time in 1925 by the Hungarian paleontologist Ferenc von Nopcsa, a renowned expert on fossil vertebrates, who recognized the existence of a new species from a few bones preserved on a fragment of shale in the collections of the Natural History Museum of Milan. Since the bones had been attributed by Italian paleontologists to an ichthyosaur, Nopcsa thought it appropriate to rename the new reptile *Askeptosaurus,* which with words borrowed from ancient Greek means "unrecognized reptile." The analysis carried out by Nopcsa on the few available bones was later confirmed by the discovery of better-preserved remains. From study of the skeleton it was thought that the askeptosaur was a lepidosaur—that is, that it belonged to the same group that includes the present-day lizards. Subsequently, however, paleontologists placed it in the order of thalattosaurs, which are archosauromorphs, relatives of the crocodiles, flying reptiles, and dinosaurs. Two other still enigmatic and hitherto unknown genera of thalattosaurs have been discovered in the Besano Formation: *Clarazia,* an approximately 1-meter-long reptile similar to a lizard but with flat teeth for crushing

At Besano there are no dinosaurs. However, the remains have been found of an arcosaur, *Ticinosuchus ferox,* which can be considered one of their ancestors. The ticinosuchus was up to 3 meters long and, like the dinosaurs, had a columnar posture of the limbs. *Drawing by F. Fogliazza.*

mollusks, and *Hescheleria,* an animal vaguely similar in shape to a salamander but of which only one very fragmentary fossil has been found.

A distant relative of thalattosaurs and placodonts seems to be *Eusaurosphargis dalsassoi,* a new, enigmatic reptile very recently described by paleontologists Stefania Nosotti and Olivier Rieppel. Its unique type specimen, completely disarticulated, reveals a peculiar anatomy, which at present does not allow experts to establish whether it was a semi-aquatic or a fully terrestrial animal.

The ticinosuchus belonged to the true archosaurs. The discovery of the first complete specimen, which was made in 1933 a few hundred meters from the town of Besano but in Swiss territory by a group of paleontologists from the University of Zurich, caused a great sensation. In fact, it was an animal very different from those previously discovered at that marine fossil site. The skeleton, about 2.5 meters long, resembled that of a crocodile. But from the position in which it was fossilized, it was plain that the reptile possessed elongated limbs extending vertically beneath the body, with the joints aligned in a non-sprawling manner. Its anatomy was certainly not that of an aquatic reptile but rather implied a highly evolved quadrupedal gait. The animal was later studied by paleontologist Richard Krebs, who named it *Ticinosuchus ferox,* "ferocious crocodile of the Ticino." Krebs recognized it as belonging to a group of archosaurian reptiles characterized by having teeth embedded in alveoli and consequently referred to as thecodonts, today considered the ancestors of the dinosaurs. This reptile probably occupied the uppermost position in the food chain on the islands and the mainland during the Italian and Swiss Middle Triassic period.

Many years after the discovery of the ticinosuchus, paleontologists realized that the structure of the forefeet and hind feet of this animal could have produced tracks similar to those of *Chirotherium,* a type of fossil footprint considered highly enigmatic for decades and found in great profusion in Permian and Triassic rocks, which outcrop in different European and Italian localities (the better-known of the latter being in Trentino and in the foothills of the Alps north of Brescia).

A last extinct group, well represented at Besano, is the protorosaurs, Triassic reptiles characterized by a more or less pronounced elongation of the neck. The relationships of the protorosaurs are still the subject of debate among paleontologists. Until quite recently, they were considered to be relatives of the lizards and were included in the lepidosauromorph diapsids; however, they are currently referred to the archosauromorph diapsids and are considered to be closer to crocodiles and the ancestors of the dinosaurs.

The Scisti Bituminosi has yielded two species that have not been found anywhere else in the world: *Macrocnemus bassanii* and *Tanystropheus longobardicus.* The former was a reptile similar in appearance and size to a lizard. It had a very delicate skeletal framework. Its neck was moderately elongated and surmounted by a small, short head with numerous pointed teeth along the edges of the jaws. Its limbs were long and slender, and the hind limbs were longer than the forelimbs. Its tail was long and thin. The most recent studies indicate that the habitat of this animal was primarily terrestrial. It may have fed on smaller reptiles, or more probably on insects. Like certain modern lizards (basilisks, chlamydosaurs), when it moved swiftly at full speed, the macrocnemus was perhaps able to balance just on its hind limbs, holding its body almost upright. The conformation of the ankle bones is, in fact, similar to that of the present-day agamid lizards, which are capable of running this way.

*Tanystropheus* is undoubtedly the most curious representative of the fauna of Besano and one of the most bizarre prehistoric reptiles in the world. It had an extremely long neck, which earned it the nickname "giraffe-reptile." In adult specimens the neck was more than half of the entire body length, which measured more than 5 meters. As in giraffes but in contrast to many long-necked reptiles such as snakes and plesiosaurs, elongation of the neck in tanystropheus was due not to an increase in the number of vertebrae (it had only eleven cervical vertebrae) but to their disproportionate development. The length of each individual vertebra reduced the flexibility of the neck (because there were only a few joints). The mobility of the neck was even further limited by the presence of peculiar ribs, which were also very long, filiform, and grouped in bundles, which constituted a sort of elastic reinforcement. At the end of the long neck, the tanystropheus had a very small skull. Younger specimens were armed with front teeth with only one point and rear teeth with three points, while in the adults the teeth were all identical and had only one point. Long, delicate limbs, the hind ones larger than the front ones, similar to those of a lizard, branched off from the short body of the animal. The tail was also very long, but shorter than the neck.

Contrasting hypotheses have been put forward about the lifestyle of this strange animal. The presence of three-pointed teeth in the younger specimens, traditionally ascribed to an insectivorous diet, in the past prompted the idea that the tanystropheus spent the early part of its life on dry land. On the other hand, nobody has noted so far that primitive pterosaurs, which were piscivorous, have multicuspidate teeth that are very similar to those of *Tanystropheus*. Other paleontologists hypothesized that teeth of this kind may even indicate a vegetarian diet based on algae, and that the tanystropheus was consequently an aquatic animal throughout its life.

The hypothesis of the aquatic habitat, which in any case seems to be the more likely, envisions a reptile that lived in a coastal marine environment and swam with undulating movements of its body and tail, boosted by its hind limbs. The somewhat rigid neck would have been held stretched forward to enable the animal to slip more easily through the water and poke its head into shoals to catch the fish and cephalopods that it fed on as an adult.

The small and large specimens of tanystropheus have always been construed as different stages of growth of individuals of the same species. However, a number of paleontologists suspect that the two forms belong to two distinct species. There are known to be different species of tanystropheus in the Middle Triassic in Germany, Spain, Romania, and Israel and in the Upper Triassic in Friuli, but no fossil site has preserved the remains of these reptiles as well as the bituminous shales of Besano. In fact, from the study of two new extraordinarily complete specimens that were uncovered at the Sasso Caldo excavation site in 1993 and 1997, the Milanese paleontologist Stefania Nosotti, associate researcher from the Natural History Museum of Milan, is obtaining new information capable of shedding more light on the anatomy and biology of these bizarre animals.

That these creatures were really strange is proved by the numerous errors that paleontologists fell into when studying their remains. One of the most resounding errors occurred with the first specimen (admittedly quite fragmentary) discovered at Besano toward the end of the nineteenth century and attributed by Francesco Bassani to a flying reptile. During the 1920s, Ferenc von Nopcsa confirmed the identification made by his Italian colleague and called the animal *Tribelesodon,* committing one of the few mistakes of his otherwise brilliant career. Shortly afterward, the discovery of numerous more complete specimens revealed that the long bones interpreted by Bassani and Nopcsa as phalanges of the finger that

With its extremely long neck, the tanystropheus, shown here capturing a squid, is the oddest representative of the fauna of Besano. *Drawing by F. Fogliazza.*

Detail of the skull of a young specimen of *Tanystropheus*. The fine bony rods extending along the side of the first two vertebrae are cervical ribs. They were grouped in bundles and gave flexibility and support to the neck. © *Natural History Museum of Milan, by courtesy of the Italian Ministry for Cultural Heritage and Activities, Superintendence of Lombardy. Photo by G. Teruzzi.*

supported the wing membrane of the pterosaurs were in fact the vertebrae of the neck, and that the general appearance of *Tribelesodon* (later renamed *Tanystropheus*) had nothing in common with that of the flying reptiles.

Today we can say that in a certain sense that error was prophetic. In fact, one century later, not far from Besano, again in Lombardy, pterosaurs would really reemerge from the Triassic. But that is another story, which deserves a chapter all its own.

# 7

# FLYING REPTILES IN ITALY

WHILE THE OCEANS of the Mesozoic were the undisputed realm of the ichthyosaurs and other marine reptiles, the absolute rulers of the skies were the pterosaurs, more commonly referred to as flying reptiles. In reality, the first animals capable of flight were insects, some of which were gigantic creatures compared to their modern kin (e.g., dragonflies with a wingspan of 70 centimeters). These creatures filled the air with humming and buzzing as early as the Paleozoic, the era before the time of the dinosaurs. Then, a few hundred million years later, at the end of the Triassic, the first vertebrates capable of flying through the air made their appearance.

It's not necessary to travel very far from where we were in the previous chapter to see them. All it takes is a 100-kilometer trip to the foothills of the Alps in the region of Bergamo. Here, the oldest flying vertebrate soared through the skies of Lombardy some 220 million years ago. Its name is *Eudimorphodon ranzii,* and it was found almost by chance in 1973 in the densely stratified calcareous outcrops in a quarry near the village of Cene in the Seriana Valley. This animal was a pterosaur, a strange reptile with membranous wings, which looked like a cross between a lizard and a bat—a sort of little winged dragon. *Eudimorphodon* was rather small: its wingspan barely reached a meter, but its descendants were soon to rule the skies, evolving into gigantic forms and preceding the birds in mastering the art of flight by at least 60 million years.[1]

## A Real Record

Once again the discovery of an important fossil site was made not by a professional scientist but by an amateur enthusiast. In 1971, Antonio Canova, a priest and teacher in Cene with a passion for geology, noted that a landslide that had occurred a few years before in an abandoned quarry had exposed a finely laminated layer of blackish rock. He picked up a few extraordinarily well preserved fossil fishes from among the rubble and hurried to report them to the Museum of Natural Sciences in Bergamo. From then on, the Cene outcrop was kept under close surveillance. More than two thousand specimens, mostly fishes and crustaceans, were collected. Then, in 1973, during an on-site inspection, the museum's preparator, Mario Pandolfi, came across a very fine slab of rock that appeared to contain long, slender bones just 1 millimeter from the surface. The slab was taken to Bergamo and cleaned up, and under a microscope it revealed its fascinating contents: the first Italian pterosaur, which was also the most ancient in the world.

Like all pterosaurs, which inhabited the planet from 220 to 65 million years ago, the flying reptile from Bergamo was able to fly due to its extremely elongated forelimbs. This elongation was a result of the disproportionate development of the phalanges of the fourth finger of the hand, which supported more than half of the wing. *Eudimorphodon* (which means "double-shaped true teeth") was characterized by no fewer than 114 teeth of different shapes and sizes (ending in one, three, or five cusps) distributed over only 6 centimeters of jaw. It had a long tail, a rather large head compared to the body, and, as can be seen from the remains of its meal (small, shiny scales found in the region of the stomach), it fed on fish, which it probably caught by skimming the surface of the water and plunging its snout into the sea once it had sighted its prey.

That's right, in the sea, because during the Upper Triassic, the present-day foothills of the Bergamascan Alps, like most of northern Italy, lay under water and formed the bottom of small, more or less deep basins within a carbonate platform.

The rocks in the Cene quarry belong to a formation called the Zorzino Limestone. These rocks formed around 210–220 million years ago in the Norian stage of the Upper Triassic by deposition of extremely fine carbonate sediment on the bottom of a marine basin more than 200 meters deep, surrounded by areas of shallow sea dotted with small islands, in a hot and rather arid climate. In this environment, the pterosaurs probably played the same part as today's marine avifauna. As attested by fossil discoveries at Cene and other localities of the valleys in the region of Bergamo (Endenna, Zogno, Berbenno) and Brescia (Valvestino), the sea was filled with crustaceans and fishes similar to those that today in-

The first Italian pterosaur, *Eudimorphodon ranzii,* was discovered in 1973 at Cene, in Seriana Valley (Bergamo). The thin slab of rock on which it is fossilized measures about 23 by 14 centimeters. Note the teeth of different shapes and sizes, which gave the animal its name. *© Photographic Archives of the Museum of Natural Sciences, Bergamo, by courtesy of the Italian Ministry for Cultural Heritage and Activities, Superintendence of Lombardy.*

habit coral reefs. Outside the water, along with the pterosaurs, lived a number of different terrestrial reptiles. These included aetosaurs, vegetarian archosaurs armored with a mosaic of bony plates that made them look rather like armadillos; phytosaurs, a group of semi-aquatic archosaurs resembling present-day crocodiles; megalancosaurs and drepanosaurs, similar to lizards and with arboreal habits, and which had a hook on the tip of the tail; and langobardisaurs, large protorosaurs related to the tanystropheus of Besano but equipped with specialized teeth for crushing mollusk shells.

After their death, these animals' carcasses were swept by currents into the marine basin and, together with a large number of fishes and invertebrates, settled on the bottom, where there were particularly favorable environmental conditions for their preservation. For some unknown reason,[2] the vertical circulation within this marine basin was so slow that in its deeper parts the water stagnated and was almost completely devoid

Three-dimensional reconstruction of the eudimorphodont, shown here capturing a fish. © *Photographic Archives of the Museum of Natural Sciences, Bergamo, by courtesy of the Italian Ministry for Cultural Heritage and Activities, Superintendence of Lombardy.*

of dissolved oxygen. Because of this, the depths were uninhabited, devoid of predatory, necrophagous organisms and aerobic bacteria, which need this gas for their vital functions. Carcasses sank to the bottom and were not devoured, and their putrefaction was also slowed by the probable acidity of the water. Sediment gradually covered them, very often before the skeletons were disarticulated, and protected them until they became fossilized.

The 1973 discovery aroused enormous interest not only in Italy but throughout Europe. The record for the oldest pterosaurs had previously been held by English paleontologists; that for the best-preserved specimen was held by their German colleagues. The rocks in the Cene quarry began to yield other remains of flying reptiles. Some belong to smaller specimens of *Eudimorphodon,* while others are referable to a different species, *Peteinosaurus zambellii.* The latter was named in 1978 by Rupert

Wild, a paleontologist from the State Museum of Natural History of Stuttgart, in honor of the paleontologist from Bergamo who had described *Eudimorphodon* five years earlier.[3]

The two or three specimens of *Peteinosaurus* (the name means "winged reptile"), all represented by incomplete skeletons or by scattered bones, were just a little larger than a pigeon (they had a 60-centimeter wingspan) and differed from *Eudimorphodon* primarily in their teeth, which were small and triangular, and lacked secondary cusps. Other differences can be found in the relative lengths of the bones of their limbs (*Peteinosaurus* had comparatively longer hind limbs and shorter wings than *Eudimorphodon*) and in their lifestyle (the peteinosaur probably fed on smaller fishes). These two species, which were found in the same fossil-bearing strata, were consequently coeval but not in competition with each other, because, as the ecologists say, they occupied two different trophic niches.

## The World's Most Ancient Pterosaurs

In 1989, another skeleton of *Eudimorphodon,* this time belonging to a young individual, was found at Ponte Giurino, also in the province of Bergamo. But Lombardy was not the only Italian region to provide science with remains of the most ancient pterosaurs. In 1982, in Carnia (northern Friuli), dark dolomite rocks of the Rio Seazza Valley, near the village of Preone, had yielded the skeleton of a new pterosaur. Here the rocks are more or less the same age as those that preserved the Lombard pterosaurs and were formed in a similar environmental context. In fact, the Dolomia di Forni Formation, chronologically very close to the Zorzino Limestone, also contains invertebrates, fishes, and other fossil reptiles belonging to the same groups found in the Upper Triassic levels of the Bergamo valleys. The discoverer of the flying reptile of Friuli, Nando Buffarini, collected the specimen during an excursion in the mountains with his wife and took it to the Natural History Museum in Udine.

Unfortunately, the skeleton of the pterosaur had already been exposed to bad weather for some time. Weakened by mud seeping into the rock, the skeleton washed away with the mud when Buffarini and his wife tried to clean the specimen. Except for a few fragments, all that remained on the surface of the small slab of limestone were the impressions of the bones. Nevertheless, by means of a silicone rubber cast, which reproduced the fossil in positive, Rupert Wild managed to study the specimen and establish the new species *Preondactylus buffarinii,* which was dedicated to its discoverer.

This first individual of *Preondactylus* (literally "finger of Preone") was the size of a magpie; it had a wingspan of less than 50 centimeters. Subsequently a second, larger individual, with an estimated wingspan of 150 centimeters, represented by three wing phalanges, was discovered in Endenna, in Lombardy. *Preondactylus* had triangular teeth with only one cusp each, arranged in a row of alternating long and short points, which do not give a clear idea of the animal's diet. Judging from its relatively longer hind limbs and shorter wings in comparison to those of other pterosaurs, Wild maintained that it was the most primitive pterosaur yet discovered. This hypothesis was based on the logical belief that the relative size of the wing should have increased during the course of evolution for the purpose of facilitating flight. For a long time, therefore, paleontologists considered *Preondactylus* to be the most basal pterosaur. However, a reexamination of Triassic flying reptiles recently carried out by Fabio Dalla Vecchia (one of the few paleontologists in the world devoted to studying pterosaurs) has shown that *Preondactylus* did not differ substantially from *Peteinosaurus.* As far as can be seen, the two pterosaurs are very similar.

Other fossils of flying reptiles were discovered in Carnia following the discovery of the first specimen. During the fieldwork for his thesis, Dalla Vecchia found a small pile of bones in the Rio Seazza Valley, which were later interpreted as the remains of a swallowed and regurgitated pterosaur. Apparently, even though pterosaurs ate fish, sometimes the opposite happened. The strata where the specimen was found abound with the fossilized remains of predatory fish, similar to the present-day garfish, which could measure up to 150 centimeters in length. Perhaps one of them caught the pterosaur as it skimmed over the surface of the water or swallowed its already lifeless body as it floated in the sea. The fish must have found the pterosaur rather indigestible, and after having chewed it, it probably regurgitated it. The remains of the unsavory meal sank to the bottom, where, once covered by the fine mud, they fossilized.

However, this was not the last discovery made in Italy. Two years later, Corrado Rosenfeld, a tenacious and successful collaborator of the Friuli Natural History Museum, discovered two specimens of *Eudimorphodon* with their characteristic teeth near the villages of Villa Santina and Forni di Sotto. One of the two fossils was nearly complete and even showed traces of the wing membrane, or patagium, in which the characteristic reinforcing fibers can be observed. On the basis of this specimen, which differed slightly from those of Bergamo, Dalla Vecchia established a new species, *Eudimorphodon rosenfeldi*. Toward the end of the 1990s, other workers from the Friuli Natural History Museum in Udine considerably increased the collection of pterosaur bones from the dark dolomite rocks of Carnia. An isolated wing phalanx was discovered by chance in a gully cut in the landslide-prone mountains that rise in the vicinity of Forni di Sopra. Even though the bone, measuring a mere 136 millimeters in length, was small, it bore witness to the presence of a true giant by the standards of Triassic pterosaurs.

## Enigmatic Creatures

Besides representing a world record in terms of antiquity, the discovery of flying reptiles in Italy has provided an opportunity to acquire new information about these very strange prehistoric animals. Many features of their skeletons remain an enigma because they are not comparable with those of any other vertebrate, due to the fact that during the course of their evolution the pterosaurs drastically modified their anatomical structure in becoming highly specialized for flight. Furthermore, since these animals left no living descendants, paleontologists can draw only vague clues from modern vertebrates to interpret the function of the various pterosaur bones. For example, we do not know with absolute certainty what function the crests on the skulls of many American species had. The most often studied case is that of the cranial crest of the pteranodon, although it is not the most ostentatious. It is hypothesized that these protuberances may have been secondary sexual characters, used for flaunting their owner's physical appearance, for ritualized combat, or for courting, or that they may have had an aerodynamic function and were used as rudders, brakes, or balancers to offset the weight of the beak. Other paleontologists believe that the crests of the pterosaurs functioned in much the same way as solar panels, for the purpose of regulating their body heat.

But the most extraordinary structures were the wings. At first glance, pterosaur wings look much more like the membranous wings of bats than the feathered wings of birds. As in chiropterans, the wing surface, supported by the forelimbs, consists of a thin integument called a patagium, which was presumably attached to the hips of the animal or farther back to the outer toes of the feet. However, in flying reptiles, as we have seen in the skeleton of *Eudimorphodon,* the skeletal structure bearing the wings is created by the elongated phalanges of the fourth finger of the hands. The first three fingers, on the contrary, are very short, are armed with claws, and protrude from the front edge of the wings, while the fifth finger has disappeared completely. In short, these animals flew by stretching out a single finger, like an extension of the arm: the finger that, on our hand, we call the ring finger. By contrast, in bats all the phalanges of the fingers are elongated. In these flying mammals, the patagium is stretched between one finger and the next, like the material between the metal ribs of an umbrella. So it can be said that bats fly with their hands open and their fingers spread.

Until a few years ago, pterosaurs were thought to be poor fliers, because the patagium supported by only one finger seemed to be a fragile structure. Recent studies have shown the exact opposite. Careful examination of exceptionally well preserved specimens with traces of the wing membranes, discovered at the famous Jurassic fossil site of Solnhofen in Bavaria, revealed that the patagium was reinforced by a dense ribbing of very fine parallel fibers, called actinofibrils, which prevented the membrane from drooping and protected it from tearing. In addition to permitting flight, the skin of the wings was a useful heat-exchange surface. In fact, since it was highly vascularized, the reptile's body temperature could be regulated by dilatation or constriction of the vessels conveying blood to it.

As in birds, the bones in the pterosaurs' skeleton were hollow in order to reduce weight. For the same reason, the thorax was reduced to a minimum, and the internal organs occupied an insignificant volume. This implies physiological adaptations that to this day paleontologists find difficult to understand. Many bony elements were fused into a single unit, so as to offer effective support for the muscles. The sacral vertebrae, for example, were welded together with the pelvic bones to form a synsacrum, which sustained the weight of the animal on landing. In the larger species, several thoracic vertebrae were fused into a similar structure, a notarium. The sternum was broad and was provided with a keel adapted for anchoring the powerful pectoral muscles that were responsible for supporting and moving the wings.

Less conspicuous, but no less intriguing, is the pteroid bone. This is a small, rod-shaped element with one end articulated to the wrist bone and the other end free but oriented toward the animal's body. This bone

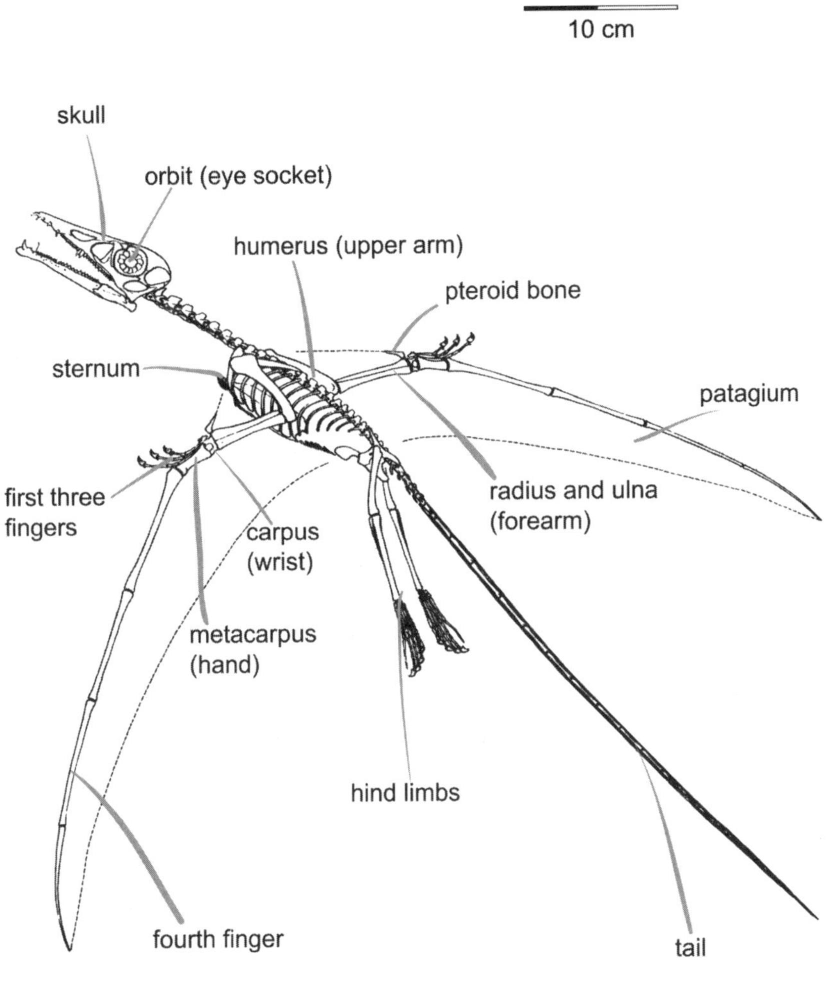


As the skeleton of *Eudimorphodon* shows, the pterosaurs flew with their forelimbs, which had evolved into wings. The hollow and very elongated bones (especially the phalanges of the fourth finger) supported a membrane of skin similar to the patagium of bats. *From Wild 1978, modified. Graphics by F. Nodo.*

is thought to have supported a portion of the patagium extending between the forelimb and the neck and to have been capable of moving the edge of the patagium slightly up and down, thereby modifying the aerodynamic profile of the anterior edge of the wing. In practice, the pteroid bone probably performed the same function as the flaps of airplanes and the alulae of birds, which help to increase lift during takeoff and landing.

During the 150 million years of their existence, pterosaurs developed an extraordinary variety of species (currently numbering more than 125). Some were not much bigger than a robin, while others reached colossal dimensions.

During the first part of their history, from the Upper Triassic to the Upper Jurassic, all the pterosaurs were characterized by a long tail, a short neck, a relatively large head always equipped with teeth, short metacarpals, and a particularly elongated outer toe. The association of all these characters typifies pterosaurs of the suborder Rhamphorhynchoidea. They were medium to small in size, no bigger than a seagull, with a wingspan of 2.5 meters at most.

A second group, the suborder Pterodactyloidea, differentiated during the Upper Jurassic from certain Rhamphorhynchoidea and superseded them completely during the Cretaceous. They reduced the tail and lengthened the neck, head, and metacarpal bones considerably, while the outer toe of the foot was reduced to a stump. Some of the more highly evolved pterosaurs were devoid of teeth and had horny bills. Other evolutionary tendencies were an increase in size, and in some groups an overgrowth of teeth and the appearance of bizarre bony crests on the skull. *Quetzalcoatlus,* whose fossilized remains date back to the Upper Cretaceous, was discovered in 1972 in Big Bend National Park, Texas. It had a wingspan of 11 to 12 meters—comparable to an F-16 fighter aircraft. This was undoubtedly the biggest creature that has ever cleaved the skies of our planet.

## And Yet They Learned to Fly

The origin of flight in these reptiles is still an enigma because no transitional forms have been found between pterosaurs and their presumed ancestors. (*Eudimorphodon ranzii* is already a true pterosaur, well specialized for powered flight, as indicated by its broad sternum and the shape of its pectoral girdle.)

## 171
## Flying Reptiles in Italy

Most specialists maintain that pterosaurs are rather closely related to the dinosaurs, with which they may even share a common ancestor: a small, agile semi-bipedal reptile called *Lagosuchus*. But according to other workers, pterosaurs were very different from dinosaurs and derived from more primitive reptiles, such as the prolacertiforms or the eosuchians, which were very similar in appearance to lizards. In the first hypothesis, the ancestors of pterosaurs probably learned to fly as a result of the need to balance themselves while running; for this purpose, besides lifting their tails, they acquired the habit of raising and spreading their forelimbs until they finally took off from the ground. In the second hypothesis, the process of mastering active flight is presumed to have begun during the Lower Triassic, or even as early as the end of the Permian, about 250 million years ago, through an initial gliding phase, which answered the need to move from one tree to another or from the trees to the ground to capture insects or flee from predators. In this view, the patagium of pterosaur ancestors probably extended between the forelimbs and the hind limbs, as in present-day flying squirrels, while the phalanges of the fourth finger and actual wings did not develop until much later.

In any case, by the Late Triassic, pterosaurs were able to fly freely and confidently through the skies. Understanding how they managed to move through the air has long been a challenge for paleontologists and aeronautical engineers. Probably their style of flight varied, depending upon their size and on the size of the wings compared to the rest of the body. In general, smaller forms, such as the Italian pterosaurs, were obliged to use more effort, while the larger forms could count on the force of the wind. The first Rhamphorhynchoidea, equipped with relatively short wings, a wide patagium, and a long tail with a diamond-shaped horny "rudder" at the tip, needed to flap their wings continuously. The more highly evolved Jurassic Rhamphorhynchoidea had longer wings and a broader patagium and consequently were more inclined to glide. The smaller Pterodactyloidea were powerful active fliers. Since they had a short tail, they had to solve a few problems of stability, but to make up for it they were more agile in performing aerial maneuvers. The large Cretaceous Pterodactyloidea with their very long wings had an aerodynamic profile very similar to that of a glider. The patagium constituted such a wide airfoil that it would have been extremely exhausting, even useless, for them to fly by flapping their wings, since it can be inferred that they were perfectly built for gliding over long distances, skillfully exploiting rising air currents.

But how did the pterosaurs move once they were on the ground? This too is an enigma that has not yet been completely solved. According to some paleontologists, flying reptiles were bipedal and digitigrade—they walked with the toes of their hind feet on the ground, like birds and theropod dinosaurs. In fact, the structure of their hind limbs is similar. Another theory, based on the fact that, at least in certain pterosaurs, the patagium was connected to the hind limbs and therefore hampered their movement, maintains that the pterosaurs were quadrupedal and plantigrade (i.e., they walked on all fours, resting the hands and soles of the feet on the ground). In fact, the fourth finger of the hand, which supported the wings, could be folded backward and upward, thereby enabling the three short taloned fingers to touch the ground. This hypothesis is supported by the existence of numerous fossil trackways (the most famous being those of Crayssac, in France) composed of strange prints of forelimbs with three fingers and hind limbs with four webbed toes.

Like birds, pterosaurs had developed a vast range of feeding specializations. By studying their teeth and snouts, it is possible to reconstruct the diets of many species. Many pterosaurs found in rocks of marine origin possessed long, pointed beaks and must have been fish-eaters. Others, of small size and equipped with peg-like teeth, were probably insectivorous. A more bizarre form, called *Pterodaustro,* which lived in South America, possessed a dense set of hundreds of long teeth as fine as needles, similar to whalebone and the thin filtering lamellae of flamingos, which must have served in the same way for capturing plankton.

It is highly likely that flying reptiles were oviparous. Although there are no known fossil pterosaur nests, from a dramatic discovery made in 1989 in Chile it can be presumed that at least one species practiced parental care and lived in colonies. A jumble of thousands of pterosaur bones from the Atacama Desert—the largest collection of these animals ever found—yielded a large number of juvenile individuals of the same species. The fossil bones were found strewn throughout ancient flood deposits, suggesting that all these specimens were caught up together in a flood about 110 million years ago. According to paleontologist Kevin Padian, this would indicate that, like many modern marine birds, pterosaurs nested in colonies on the ground. Possibly the young individuals had a prolonged nesting stage during which their parents cared for them until they were grown enough to fly on their own.

One of the mysteries yet to be solved concerns their metabolism.

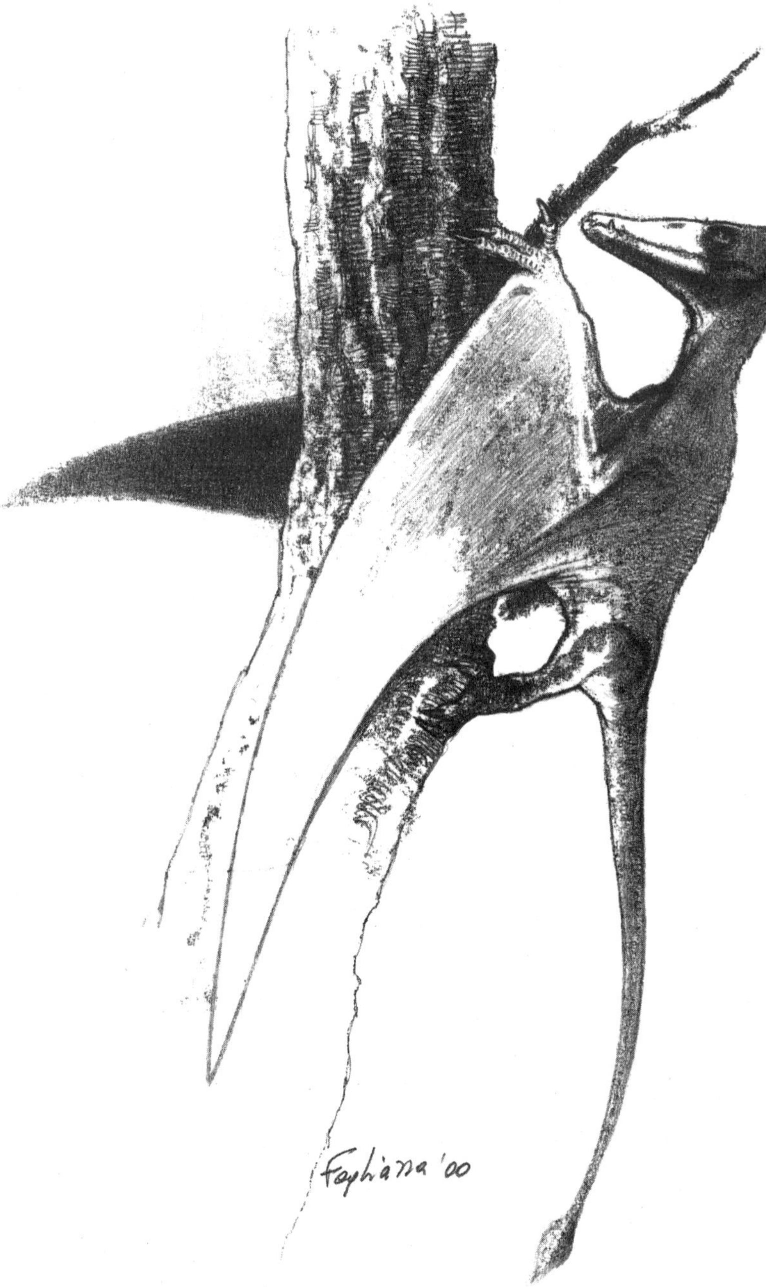


*Eudimorphodon,* which lived in northern Italy about 220 million years ago, is considered to be the world's most ancient pterosaur. The long tail, shared in common by all the flying reptiles of the suborder Rhamphorhynchoidea, is a character of primitiveness. *Drawing by F. Fogliazza.*

How did reptiles (usually considered to be cold-blooded animals) maintain the activity necessary to enable them to actively fly and even flap their wings? This question still has no universally accepted answer. First and foremost, it may be that pterosaurs were not cold-blooded animals. After all, the vertebrates that fly in modern skies (birds and bats) are warm-blooded. Flying involves the use of energy that must be immediately recovered, and an animal that would have to sit still in the sun for hours before it was able to fly again is inconceivable.

It has been hypothesized that the pterosaur's body was covered with fur, with the exception of the wing membranes, to prevent loss of heat. This hypothesis is based on the discovery in Kazakhstan of the skeleton of a pterosaur called *Sordes pilosus* ("hairy demon") by the Russian zoologist Sharov in 1971. Sharov describes the specimen as apparently surrounded by thick hair. This would appear to prove that flying reptiles needed to maintain a high body temperature—in other words, that they had a metabolism similar to that of birds and mammals. Recently David Unwin, of the Museum of Natural History in Berlin, together with his wife Natasha Bakhurina, restudied the specimen and affirmed that most of what was thought to be hair was in actual fact the traces left by frayed fibers of the patagium.

But in 2002, Chinese paleontologists found a second, better-preserved, and clearly "haired" pterosaur fossil in the Inner Mongolia Autonomous Region in North China, providing new evidence that at least some flying reptiles were covered by a sort of insulating integument.

In any case, it is certain that pterosaurs were a successful group. They shared their existence with the large aquatic reptiles and the dinosaurs, the giant rulers of the Mesozoic lands, but with them they also shared the tragic fate of extinction. Some 65 million years ago, they disappeared forever, without leaving descendants, taking with them their numerous mysteries.

The first clues to that mass extinction of planetary proportions were found a few dozen years ago, once again in Italy, across the Apennines near a delightful medieval village called Gubbio.

# 8

# GUBBIO: THE ECHO OF DISASTER

SIXTY-FIVE MILLION YEARS ago, an event that would be difficult for a human being to imagine wiped out the dinosaurs forever. The dinosaurs were not the only victims of the devastation. About 75 percent of the then-living species of animals and plants (including large marine and flying reptiles, and whole groups of invertebrates such as ammonites) were eliminated. The groups that did survive suffered severe losses.

The end-Cretaceous extinction was not the only mass extinction that has stricken our planet, just the best known. At the end of the Permian Period, for example, there was an even more catastrophic event, which led to the disappearance of more than 90 percent of all living creatures. According to analyses carried out by various paleontologists, disastrous biological crises take place at periodic intervals and are determined by relatively rapid changes in the earth's global climate.

The enigma of the disappearance of the dinosaurs has fascinated generations of scholars, who have advanced dozens of hypotheses to explain it. But the one given the most credence originated and gained ground right in Italy, in a splendid little town in Umbria. Gubbio is renowned throughout the world for its stone houses, towers, and palaces that look out onto medieval streets imbued with timeless charm. The feeling of peace and serenity that it inspires in visitors makes it extremely difficult to believe that evidence of an incredible planetary catastrophe exists at this very spot.

## The Archives of the Earth's History

Gola del Bottaccione, near Gubbio, is regarded by geologists as an extraordinary archive for studying the history of the earth. The walls of this spectacular canyon are composed of a succession of sedimentary rocks formed from deposits of calcareous ooze on the bottom of the Tethys, the ancient ocean that once covered the entire area. The extraordinary thing about these layers is that the sedimentation that created them was continuous, without any interruption, over a period ranging from 140 to 50 million years before the present. Then, about 15 million years ago, when a portion of the seabed was heaved up to form the Apennine chain, these rocks emerged and the strata were cleaved by the erosion of river waters.

During the 1970s, the American geologist Walter Alvarez, son of the better-known Luis Alvarez (who received a Nobel Prize in physics in 1968), happened to be in this area, together with the geophysicist William Lowrie, to study the earth's geomagnetic field reversal. Certain rocks contain grains of metallic minerals capable of recording the direction of the earth's magnetic field at the time the minerals were formed. They acted like "fossil compasses," from which paleomagnetics experts are able to obtain information about the movements of the continents during the geologic eras and about the formation of mountain ranges.

Gola del Bottaccione is renowned for its extensive outcrops of Scaglia Rossa limestone, which contains hematite (an iron oxide capable of recording the earth's magnetic field), so Gubbio was an excellent place for the two American scientists to collect samples. During their field trips, they were sometimes accompanied by Isabella Premoli Silva, a micropaleontologist who had used the Scaglia from around Gubbio for her studies on foraminiferans, small unicellular organisms that constituted a significant part of the plankton living in the ocean at that time, whose shells contribute to the formation of limestone throughout the world. Premoli Silva showed Alvarez and Lowrie the stratigraphic boundary between the Cretaceous and Tertiary (abbreviated as "K/T boundary" by scientists), which in the rocks at Gubbio was characterized by an approximately 1-centimeter-thick layer of clay, completely sterile and devoid of fossils. While the layers dating back to the Cretaceous contained an abundant and diversified population of plankton, with foraminiferans of the genus *Globotruncana* represented by large specimens visible with the naked eye, they disappear completely in the K/T boundary and then reappear in subsequent rocks corresponding to the Tertiary period with minute specimens belonging to the species *Globigerina eugubina*. It was the layer of clay completely devoid of fossils that most impressed and totally diverted the attention of Alvarez. He began to forget problems

The K/T (Cretaceous-Tertiary) limit as it appears in the rock bedding of the Gola del Bottaccione, near Gubbio. Level G 97 B belongs to the last stage of the Cretaceous (Maastrichtian), while G 97 C was deposited at the beginning of the Tertiary (Paleocene). Between the two levels, 98 centimeters along the tape measure, lies the layer of clay devoid of fossils that attracted Alvarez's attention. *From Cita and Premoli Silva 1974.*

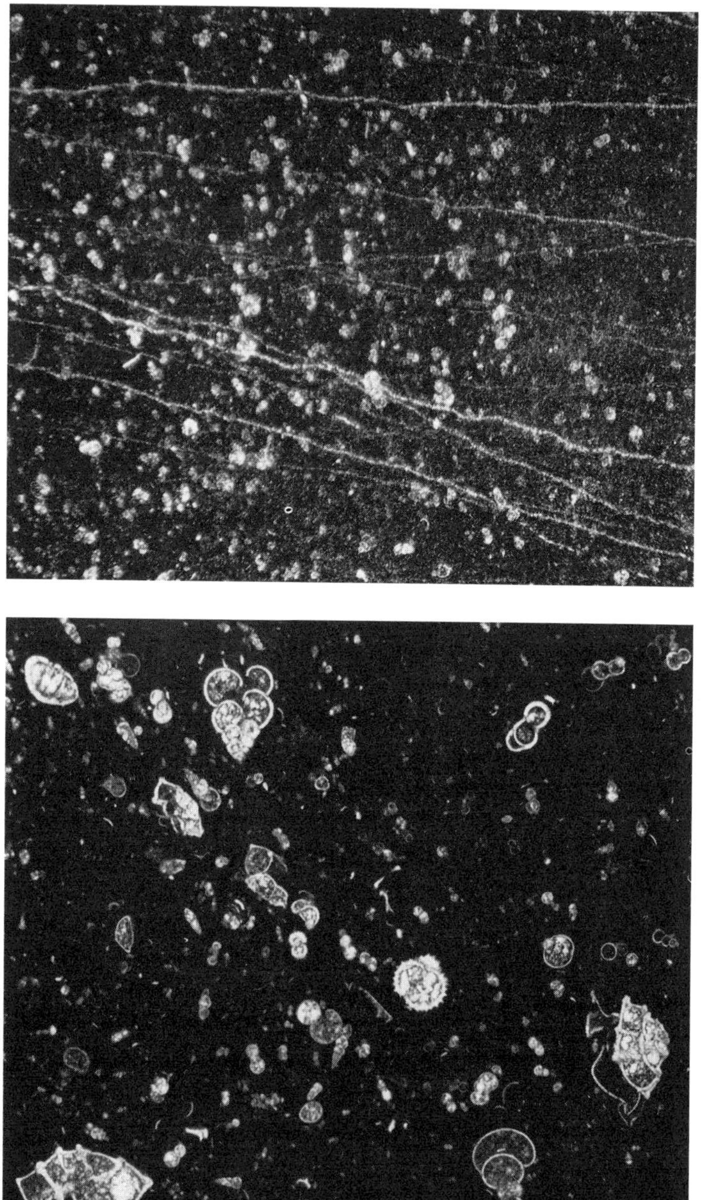

Observed under the microscope (at the same magnification), the levels G 97 C (*top*) and G 97 B (*bottom*) show evident differences in microfossil content. The Tertiary foraminifers are much smaller and show a minor variety of species compared to those of the Cretaceous, which in the upper level disappeared completely. *From Cita and Premoli Silva 1974.*

of paleomagnetism and to ask himself why the foraminiferans had suddenly disappeared in that way. What had happened? What did that sterile clay mean?

Everything pointed to one conclusion: that the disappearance of these microfossils must have been sudden and catastrophic. Another disconcerting fact was that their extinction, recorded in the K/T boundary of the Gubbio limestone, dated back to more or less 65 million years ago, and thus to the same period that had marked the end of the reign of the dinosaurs on earth. This coincidence would lead Walter Alvarez to formulate one of the most illuminating and at the same time controversial theories in the scientific world.

## An Extraterrestrial Cause

In 1978, working with his father and two nuclear chemistry experts, Frank Asaro and Helen V. Michel, Walter Alvarez conducted some sophisticated chemical analyses (neutron activation analysis) on samples of rock from the Gola del Bottaccione. The results were astounding. The layer of clay devoid of fossils recorded a concentration of iridium at least thirty times higher than in the underlying or overlying sediments. Since iridium is a metal found in very limited quantities in terrestrial rocks but contained in greater percentages in meteorites, the scientists concluded that this element had extraterrestrial origins, and that this had to have something to do with the widespread extinctions of the Cretaceous. However, before working out any theory, it was of paramount importance to know whether that anomalous concentration of iridium characterized only the Gola del Bottaccione or whether it was typical of the K/T boundary in other parts of the world.

When Walter Alvarez came across the clay in Gubbio, there was only one other site in Europe in which the K/T boundary was known to be exposed. This was the Stevns Klint reef, south of Copenhagen. Here, too, a layer of clay separates limestone of the Maastrichtian from that of the Danian (the last stage of the Cretaceous and the first stage of the Tertiary, respectively). It also has an anomalous concentration of iridium.

So where did all that iridium come from? One of the first possibilities that Alvarez and his colleagues considered was a supernova, a hypothesis that had previously been proffered by paleontologist Dale Russell and physicist Wallace Tucker. According to these two scientists, an exploding star (supernova) relatively close by could have scattered iridium onto our planet, while the radiation produced by the explosion would undoubtedly have been devastating for the living creatures of the earth.

How could a hypothesis of this kind be confirmed? If the cause of the mass extinction of the dinosaurs and many other species on our planet really had been a supernova, analysis of the K/T clay should yield traces of plutonium-244, a radioactive isotope formed by the nuclear reactions of stellar explosions. This element generally decays with a half-life of 83 million years; since "only" 65 million years had elapsed following the end of the Cretaceous, analysis should have been able to reveal its presence. The samples of Gubbio clay, however, yielded negative results. There was no trace of plutonium-244. This meant only one thing: there had been no supernova explosions close to the earth during the period in question.

## The Impact

At this point, it was necessary to come up with a new hypothesis to account for the extinction of the foraminiferans (which also coincided with the extinction of the dinosaurs) and the presence of the iridium. Perhaps the explanation was an impact with a gigantic asteroid. But how could it have caused the annihilation of so many living species on a worldwide scale? To be able to answer this, it was necessary to hypothesize a terrifying scenario—something truly apocalyptic, capable of killing or decimating plants and animals even in the hemisphere opposite the one in which the catastrophe occurred. What could have happened on that far-off day 65 million years ago?

A cataclysm of this kind could be explained only if an immense meteorite at least 10 kilometers in diameter had hit the earth at the incredible speed of 30 kilometers a second. Its entry into the atmosphere would have been devastating, because with its kinetic energy, the celestial body would have had a destructive power equivalent to 100 million hydrogen bombs. The result would have been a blinding flash of light and a deafening blast, followed in the space of a few seconds by an immense shock wave right down into the bowels of the earth and by incandescent vapors and deadly crossfire as a myriad of projectiles of terrestrial and extraterrestrial matter were hurled into the air over great distances. At the point of impact, rocks would have instantly melted and even vaporized. This would have been an almost unconceivable physical event, one of unimaginably terrifying violence. All forms of life would have disintegrated in a few seconds within a radius of hundreds of kilometers. That day, no creature would have survived the event, even at a distance of several thousand kilometers.

The impact would have blasted such an enormous quantity of dust into the air as to plunge the entire planet into darkness. The earth would have remained dark and cold for several months, until most of the dust raised by the impact had settled out. The temperature would have dropped by many degrees, rendering the planet inhospitable, while, deprived of sunlight, plants would no longer have been able to grow, causing an interruption in the food chain. This would have caused the death of all the herbivorous animals and then the carnivores that fed on them, resulting in mass extinction.

This hypothesis, of course, prompted great controversy among scientists, with heated debate between those in favor and those against the idea put forward by Alvarez's team. Many were skeptical and clung firmly to more classical viewpoints that contemplated the gradual disappearance of the extinct species. Many scientists were of the opinion (and this remains one of the greatest objections to the theory of the meteoritic impact) that the extinction of many living species was caused by changes in important environmental factors which were assumed to have slowly taken place during the Cretaceous: retreat of shallow seas from low-lying areas of the continents, continental breakup, and a series of climatic changes that proved detrimental to the dinosaurs.

Even so, the impact theory seemed convincing and gained new supporters, although in actual fact it still lacked confirmation. An asteroid of such destructive power must inevitably have left its mark. So where was the crater produced by an event of such proportions? According to estimates, it must have had a diameter of at least 150–200 kilometers. Although one would expect a feature of that size to be pretty obvious, it took many years to find it.

## The Evidence

As soon as Walter Alvarez, his father Luis, and their collaborators discovered the anomaly in the concentration of iridium at Gubbio, numerous pieces of evidence in favor of the impact theory began to accumulate. Besides Gubbio and Stevns Klint, K/T sites in various parts of the world soon numbered more than 100, and they all showed an anomalous quantity of this metal. Some localities also yielded so-called shocked quartz and microtektites right at the K/T boundary. The former are grains of quartz deformed by pressure from a high-speed shock wave. Grains of quartz of this type had previously been found near known meteoritic craters and in nuclear testing areas. Tektites are glass spherules formed by rapid cooling of blebs of liquid rock melted and blasted into the air by a violent impact, which subsequently fall to earth. One of the first to recognize the importance of these spherules was Alessandro Montanari, a geologist from Umbria who played a very important part in Alvarez's study team.

While these discoveries continued to sketch out an increasingly detailed picture of the dynamics of the catastrophe and its planetary scale, they still lacked the most important evidence: the crater.

Anyone who has attentively observed the moon with the help of a telescope or even with binoculars knows that it is pitted with craters caused by meteorites. They date back to early in the formation of our solar system, when the planets and their satellites were subjected to almost

Meteor Crater in Arizona is one of the few meteorite-impact craters visible on earth today. It has a diameter of 1 kilometer and 200 meters and a depth of 210 meters and was formed about 50,000 years ago by the impact of a small asteroid that measured only 45 meters in diameter. *Photo by C. Dal Sasso.*

continuous bombardment by debris drifting in space. Those enormous scars have remained exposed because the moon still retains its original surface. In contrast, on the earth, craters caused by impacts have been effaced by erosion or covered with sediment or vegetation. Others may be located on the seabed or beneath the frozen polar caps. Last, many may have been subducted—swept down and melted into liquid rock beneath the plates of the earth's crust, where the plates collide with one another to form oceanic trenches and mountain ranges. As a result of tectonic movements, the earth's crust is slowly but continuously recycled (the crust of all the present-day oceans was formed during the last 70 million years). It could be said that, unlike the moon, the earth "changes skin."

All this made the search for the fateful crater a feat somewhat like finding the proverbial needle in a haystack. Despite the difficulties, at the beginning of the 1990s investigations finally met with success. The scar left by the collision was found at Chicxulub, in the Yucatan peninsula in Mexico. This vast crater has a diameter of almost 200 kilometers and is largely filled with sediment that has been deposited over the 65 million years following the cataclysm. Located mostly on the bottom of the Gulf of Mexico (the portion above sea level is now covered by a lush tropical forest), the site had already been discovered at the end of the 1950s by geologists employed by the Mexican oil company Pemex. During gravimetric surveys, they had found an enormous circular structure. Once drilled, however, it proved to be totally lacking in oil and was thought to be a buried volcano. Years later, in 1981, geophysicists Antonio Camargo and Glen Penfield, who worked for the same oil company, began to suspect that the structure might have been caused by an impact, but they did not announce the results of their research to the scientific world.

## The Weak Point

Despite the tremendous ravages envisaged by the scenario of the impact theory, it was true that many plants and many animals (insects, birds, and mammals as well as many reptiles such as turtles, crocodiles, and snakes) survived the crisis at the end of the Cretaceous. This is undoubtedly one of the most bewildering questions: Why did an event of such proportions exterminate thousands of species and allow others to survive? How can a planetary catastrophe operate selectively?

According to French paleontologist Eric Buffetaut, the explanation is linked to the different availability of food resources. As we mentioned previously, the impact released an enormous quantity of dust into the atmosphere, which acted as a filter to the sun's rays, hindering the normal process of photosynthesis, which drastically diminished plant life. The first animals to die out were herbivorous dinosaurs, followed by the predators that fed on them. This crisis presumably did not cause particular damage to animals that lived in fresh water (fishes, amphibians, turtles, and crocodiles), whose food chain was based on detrital organic matter. In fact, rather than decreasing their food supply, devastation of biological communities on land would have greatly increased the abundance of organic matter washing into aquatic systems. In the same way, worms and many insects that fed on decomposing matter in the soil were unaffected. Thanks to this unexpected source of food, lizards, tortoises, and small mammals also managed to survive the crisis.

The plants that survived were those with the hardest and toughest seeds and those capable of adapting to the harshest weather conditions, similar to the plant associations that today populate the arctic tundra.

## A Jumble of Hypotheses

Of course, not everyone is in agreement with the meteorite theory or with the hypotheses that from time to time have been advanced to complement it. According to some experts, the abnormal concentration of iridium may not be of extraterrestrial origin.

In 1983, a group of American volcanologists managed to measure the quantity of iridium present in gases emitted during the eruption of the volcano Kilauea in the Hawaiian Islands and found the concentrations very high. These experts maintained that in all probability, matter so rich in iridium could come from the lowermost zone of the mantle, the intermediate part of the earth that extends deep beneath the crust. In fact, iridium, which is one of the six elements in the platinum group, is siderophilous—it is attracted by molten iron—and therefore, during the first stages in the formation of our planet, it followed this metal, which together with nickel went on to form the heavy core at the center of the earth. This is also the reason why it is extremely rare on the earth's surface. According to the volcanologists, once the iridium had been emitted in the gaseous state from the volcanic craters during eruptions, and after dispersing into the atmosphere, it may have been deposited on the earth's surface in large quantities during periods of particularly active volcanism. Therefore, it is possible that the anomaly in the K/T boundary clays was caused by intense and prolonged Late Mesozoic volcanic activity, not by the impact of a large asteroid.

In attempting to reconstruct a scenario of this kind, several scientists, among them Vincent Courtillot of the University of Paris, focused their attention on violent volcanic eruptions that occurred precisely 65 million years ago in western and central India. These events occurred over an exceptionally long period of time, on the scale of millions of years, and undoubtedly had devastating effects, given that the lava eventually covered an area of approximately 500,000 square kilometers. The resulting accumulation is known as Deccan traps, basaltic lava flows hundreds of meters thick that are exposed in the Deccan region of India. During the tremendous eruptive phases, the atmosphere must have been so heavily laden with clouds of dust and volcanic gases as to create enormous disturbances in the terrestrial ecosystems, which, depending upon the hypothesis, involve different scenarios. Some workers think the atmospheric changes caused a global drop in temperature. The increasingly bitter cold weather made the survival of life on earth extremely difficult. Others, on the contrary, maintain the exact opposite: that the climate grew hotter. The air became filled with carbon dioxide, giving rise to a hothouse effect capable of raising global temperatures by at least 10 degrees. This was accompanied by tremendous acid rainstorms caused by changes in the composition of the atmosphere, which destroyed most of the forests and rendered the surface waters of seas and oceans acid and inhospitable even to a depth of several meters. Hence the inevitable alteration in the food chain that ended with the extinction of many animal species, including the dinosaurs.

## The Dinosauroid

The mystery of the disappearance of the dinosaurs has whetted the curiosity of hosts of paleontologists and other scientists. Over the years, they have put forward dozens of theories, some very serious, others less so, and some downright crazy. It was suggested, for example, that the widespread extinction of the Cretaceous Period occurred because of sudden epidemics, because of a malfunction of the pituitary gland that caused excessive growth of the bones and other disfiguring elements such as horns, plates, and spines. It was also proposed that small mammals became specialized for devouring dinosaur eggs or that the huge reptiles suc-

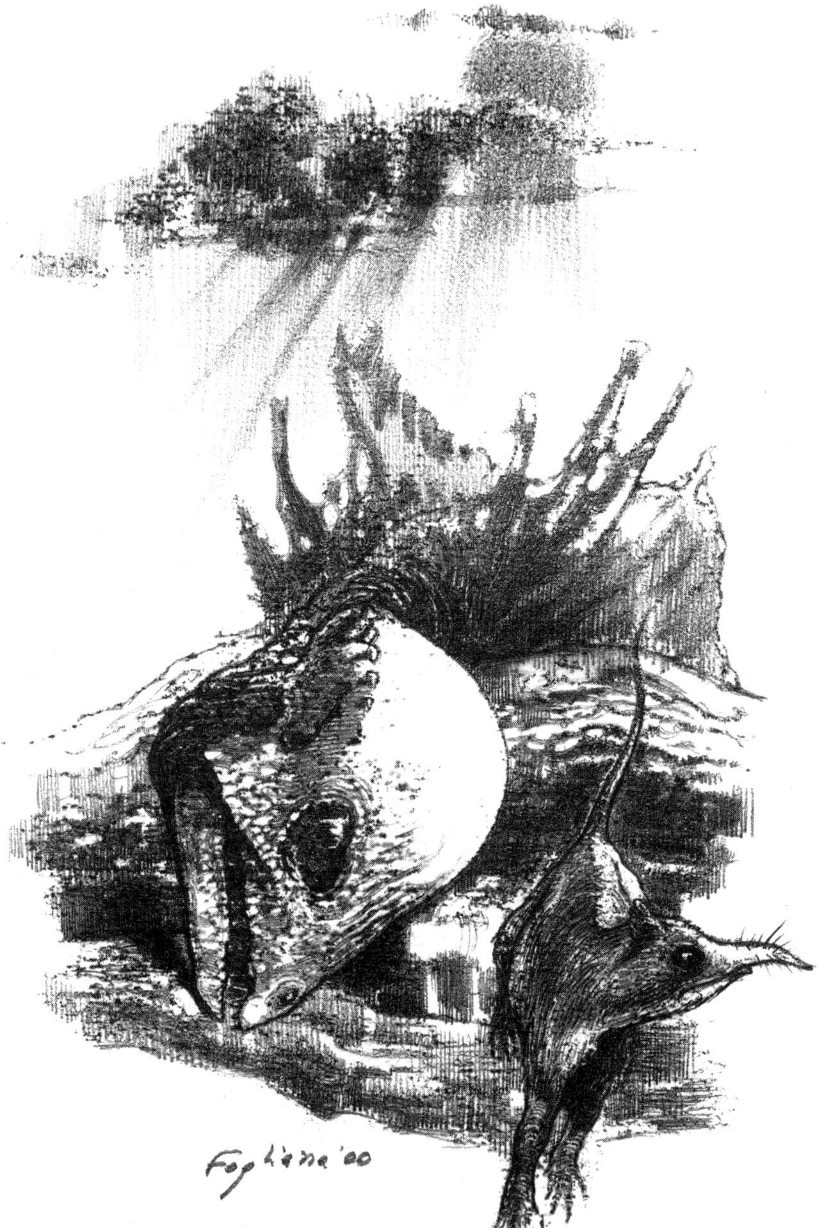

The mass extinction of the great reptiles at the end of the Cretaceous aided the rise of the mammals, which until then had survived as rat-like forms that led a primarily nocturnal life. *Drawing by F. Fogliazza.*

cumbed to mental diseases due to an excessive reduction in the size of the brain. Some paleontologists suggested a crisis in the food chain due to "changes in the vegetation." But in actuality, the appearance and diffusion of angiosperms (flowering plants), which began in the Lower Cretaceous, if anything may have encouraged the evolution of new groups of herbivorous dinosaurs, such as the ceratopsids and the hadrosaurs. The humorist Will Cuppy solved the problem with a witty remark. In the final analysis, he said, those huge, lumbering, dull-witted beasts were an evolutionary error that had lasted too long. Someone went so far as to suggest, with little good intellectual taste, that the Martians had something to do with it.

In one way or another, the dinosaurs disappeared. On this point the fossil evidence is clear: there is no trace whatsoever of dinosaur bones in rocks less than 65 million years old anywhere in the world. On the contrary, those rocks are rich in bones belonging to another group of vertebrates: the mammals.

*Mors tua, vita mea.* There are many who believe that if it had not been for the disappearance of the great Mesozoic reptiles, human beings would not be here. It was precisely from that moment on that the primordial mammals of the Cretaceous, previously obliged to maintain small dimensions, a rat-like appearance, and a nocturnal lifestyle, could finally come into the open and express all their evolutionary potential. Mammals inherited the earth and all the environments left free by the reptiles, reoccupying ecological niches in their own way and diversifying into the myriad of species we know today. This evolutionary process took dozens of millions of years and in the end led to humans, who were destined, among other things, to become the only animal capable of wittingly contributing to its own extinction.

But what would have happened if the dinosaurs had not disappeared? Canadian paleontologist Dale Russell is convinced that they would have become creatures almost as intelligent as we are. Spurred by a creative impulse, Russell built a model of an evolved dinosaur in fiberglass with the help of taxidermist Ron Sequin. In Russell's vision, such human-looking dinosaurs would have warm blood, an erect posture, a bipedal gait, and a body covered with scales; the reptile tail would have disappeared. They would have fed their young for a long time by regurgitating pre-chewed food. Their large, round skulls would have allowed them some form of evolved language, and their lives would have been spent mainly in groups of socially organized hunters.

Russell's notions were based on the discovery he made in 1968 in the Canadian Badlands of Dinosaur Provincial Park, Alberta, of a carnivorous dinosaur weighing about 40 kilograms and just over 120 centimeters long. The size of the cranium clearly indicated that the *Stenonychosaurus* (now synonym of *Troodon*) had a rather large brain with an encephalization quotient that made it more intelligent than most of the dinosaurs and mammals belonging to the same period. According to Russell, if the evolution of this dinosaur had been able to continue, it would have developed the innermost digit of the forelimb, which would have given it greater ability in manipulating objects. If not for the mass extinction that hit the dinosaurs, the evolutionary tendency of these theropods would have gone in a "humanoid" direction and led to more and more intelligent forms. This idea was strengthened by the discovery in 1985 of the first complete *Troodon* skull. This animal effectively had a brain three times larger than other reptiles of comparable size known up until that time. The ratio between the weight of its entire body and its cerebral mass was similar to that of a modern ostrich.

In short, if 65 million years ago the dinosaurs had not disappeared forever, not only would we not be here, but in all probability, as Dale Russell would put it, this book would have been written by one of them.

## Not All Extinct

Science fiction apart, in reality there are growing numbers of scientists willing to admit that at least some types of dinosaurs left descendants. This is a new way of looking at evolution, which can be summed up in a phrase by the famous paleontologist Robert Bakker: "The dinosaurs are still alive today, but we call them birds."

The first evidence of a close link between these two groups of vertebrates was the discovery in 1861 in a locality in Bavaria (Solnhofen), Germany, of *Archaeopteryx,* a fossil dating back to 165 million years ago that then became famous throughout the world. Had it not been for the presence of fossilized feathers around the bones of four of the seven specimens discovered to date, *Archaeopteryx* would not have been classified as a bird. In fact, many of its features are reptilian: jaws with teeth, a small, non-carinated sternum, forelimbs with three functional fingers, a typical dinosaurian pelvic girdle, and a long, bony tail. But its body was almost entirely covered with feathers, which undoubtedly served the purpose of providing very efficient thermal insulation. The forelimbs were furnished with large flight feathers (remiges) and long tail feathers (rectrices). This animal, about the size of a pheasant, was undoubtedly able to fly, but it probably chased and captured insects and small reptiles on the ground.

Since the discovery of *Archaeopteryx,* the hypothesis of a direct evolutionary link between dinosaurs and birds has gained enormous support, despite the opinions of those who thought that these groups merely had a common ancestor among the reptiles and that all the resemblances observed in their skeletons were the outcome of adaptive convergence. In fact, during the decades following 1861, there were a succession of discoveries of carnivorous dinosaurs that revealed a skeletal anatomy remarkably similar to that of the fossil found in Bavaria as well as birds more generally. During the 1920s, *Velociraptor,* which had long forelimbs and a pectoral girdle with a furcula and sternal plates, and *Oviraptor,* which had similar features as well as a horny beak resembling that of a parrot, were unearthed in the Gobi Desert from sandy beds of the Upper Cretaceous. A specimen of *Oviraptor* was later found buried with its fifteen eggs in the posture typical of a brooding animal.

However, it was not until 1995 that the evidence in favor of the hypothesis that birds evolved from dinosaurs became almost incontestable. Near the village of Sihetun (Liaoning province, in the People's Republic of China), a farmer extracted a slab of rock containing the fossil remains of a small carnivorous dinosaur. Finding it rather strange, he sent it to paleontologists at the National Geological Museum of Peking, who immediately realized the extraordinary value of the discovery: the dinosaur's body was covered with feathers. *Sinosauropteryx,* as it was later called, was the first of an incredible variety of fossils that were shortly to emerge from the quarries of Sihetun, all exceptionally well preserved in a natural casket of lacustrine sediments deposited between the end of the Jurassic and the beginning of the Cretaceous about 120 million years ago. Over a span of three years, in addition to this *Sinosauropteryx,* at least four other feathered and plumed dinosaurs, called *Protarchaeopteryx, Caudipteryx, Beipiaosaurus,* and *Sinornithosaurus,* were discovered.

While the Italian *Scipionyx* is still the best source of information for studying the internal anatomy of these reptiles, the feathered dinosaurs of China are true "missing links." They shed light on the most fascinating evolutionary transition in the history of vertebrates: the origin of birds

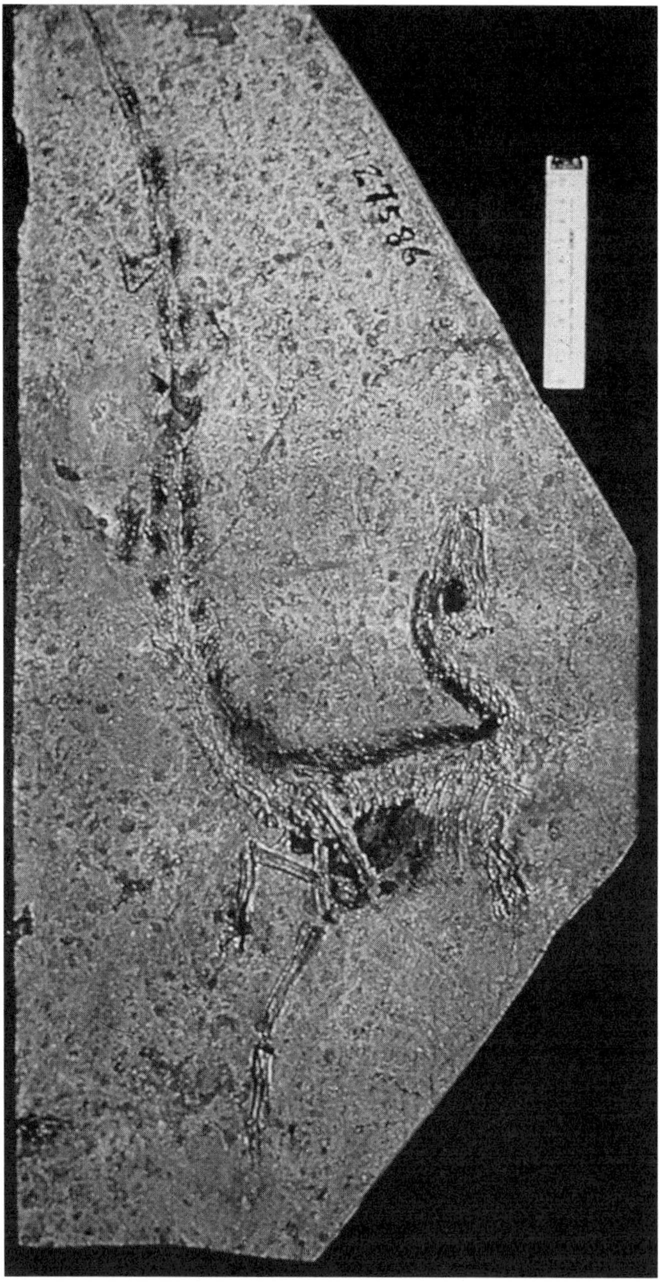

This specimen of *Sinosauropteryx,* a small coelurosaur discovered in 1995 in the Chinese province of Liaoning, radically changed our ideas about dinosaurs. As can be seen in this exceptional photograph, the reptile's neck, body, and tail are covered with blackish filaments similar to bird feathers. *© Institute of Vertebrate Paleontology and Paleoanthropology, Beijing. Photo by Z. Dong.*

through the gradual appearance of filaments, plumules, protofeathers, and feathers on the bodies of carnivorous dinosaurs similar to *Velociraptor.* The dinosaurs of Sihetun are bizarre creatures. They could not fly, because their forelimbs were too short and were armed with predatory talons, but they possessed all those "preadaptations" that enabled some of their relatives to take wing.

Birds are consequently direct descendants of coelurosaurian theropod dinosaurs. Since dinosaurs that lived in trees are not known, it would appear that in the coelurosaurs, the ability to fly developed directly on the ground and not as a result of an arboreal life.[1] Initially the covering of scales became transformed into rudimentary feathers to give the little dino-birds greater thermal insulation. The tips of the forelimbs then developed longer feathers, which served as stabilizing appendages to help the animals keep their balance when leaping to capture their prey. At some point, after the animals acquired the ability to fly, improvement of these structures was guided by requirements of an exclusively aerodynamic nature. *Archaeopteryx*'s primary and secondary flight feathers have the same shape and the same disposition on the wing as those of modern birds. Consequently, it is highly unlikely that the selection that produced this conformation was guided by factors unrelated to the dynamics of flight.

Feathers are the most unmistakable anatomical feature of birds. These structures are the ideal solution for flight. They are lightweight and efficient, and unlike the patagium of pterosaurs and bats, they can be replaced whenever they become damaged.

Like the carnivorous dinosaurs they derive from, birds are bipedal. After freeing themselves first of the ambulatory function and then of the predatory function, the forelimbs became specialized into articulated wings. The skeletal peculiarities of birds are related to the development of the powerful pectoral muscles, which permit active flight. The sternum is highly developed, with a prominent muscle insertion area referred to as the keel. The clavicles, which are also connected to the sternum, are fused into a Y-shaped bone (the famous furcula). In order to be able to take off from the ground, birds' bodies have become very much lighter due to reduction in the thickness of the bones, which are also hollow, and also to the development of large air sacs connected to the lungs. In the cranium, the large cerebral volume is related to the need to coordinate the complex movements that make flight possible, while the horny beak represents an alimentary adaptation.

The close relationship between dinosaurs and birds has now been demonstrated by the numerous fossils of "feathered" theropods discovered in China: portrayed here are the compsognathid *Sinosauropteryx* (*bottom left*), the dromaeosaurid *Sinornithosaurus* (*left*), and the therizinosaurid *Beipiaosaurus* (*right*). Above them are several primitive birds of the genus *Confuciusornis*. *Drawing by Luis Rey.*

In its general structure, however, the skeleton of a bird is basically that of a theropod: chickens, turkeys, and ducks, the birds we eat, as well as all those that delight us with their cheerful twittering and multicolored plumage, are the true descendants of dinosaurs.

# 9

# THE FUTURE OF ITALIAN DINOSAURS

## Dinosaurs of the Third Millennium

THE DRAGONS OF POPULAR legends, which have existed in people's imaginations since time immemorial, have now been replaced in the minds of adults and children by dinosaurs. Dinosaurs are "modern dragons" in every respect. They are no longer monsters created by fantasy but creatures whose actual existence, as has been more than demonstrated in this book, is proved by scientific evidence. Although these dragons became totally extinct (except for their avian descendants) millions of year ago, they have now been resuscitated by studies carried out by paleontologists aided by futuristic computer-graphics technologies.

Paleontology is currently experiencing a true Golden Age. Over the last twenty-five years, new dinosaurs have been described at a rate of one every two months, doubling the number of known species. And now, thanks to the speed at which information travels since the advent of Internet, announcement of these discoveries has reached an incredible average of one dinosaur every three weeks.

There are at least three times as many paleontologists devoted to studying dinosaurs now as there were sixty years ago. More than half of the approximately nine hundred species of dinosaurs that we know today were unknown in 1970. Dinosaurs have been discovered on all the continents, and even seemingly better-explored countries—the case of Italy is emblematic—continue to offer wholly unexpected discoveries, providing such extraordinary information as to radically change the image we have of dinosaurs.

However, the dinosaurs of the third millennium will not materialize overnight. From estimates based on the diversity of present-day vertebrates, that of dinosaurs is expected to be in the neighborhood of fifteen hundred to two thousand species, which means that there are still a lot of finds to be made.

As a result of this rejuvenation of the field, and thanks also to computer graphics with hitherto unimaginable evocative powers, paleontology is attracting the attention of an increasingly large and well-informed audience. If Crichton and Spielberg had not invented *Jurassic Park,* perhaps this newfound passion for dinosaurs would never have attained the proportion of a fashion craze, and as important as they may be from a scientific standpoint, "Ciro" & Co. would probably not have become as famous as they are in Italy and throughout the world.

## And Now What?

Even an up-to-date summary such as this book is soon likely to become dated. Within a few years, there will be new discoveries that will enable us to further our knowledge of the dinosaurs that existed in this part of the world. What discoveries lie in store for us? What can we reasonably expect?

We must first recognize that the history of Italian dinosaurs started out rather well: the three kinds recognized to date (*Scipionyx,* "Antonio," and the saltriosaur), thanks to the discovery of their skeletal remains, are the representatives of three brand-new species. Even though only one of them has been officially described, it can confidently be said that these dinosaurs are endemic forms that are unlikely to be found anywhere else. They are genuine "made in Italy" dinosaurs, with authenticated documentation of origin. Most probably, the reason for this uniqueness is the repeated isolation of the carbonate platforms that constituted paleo-Italy in the Mesozoic—that broken-up archipelago of islands and peninsulas scattered across the Tethys Sea. Another peculiarity of the Italian dinosaurs is that they all were found in sedimentary rocks of marine origin. Imagining the future of studies on Italian dinosaurs, it is therefore logical first to presume that other new species will be discovered in these rocks.

Another factor that makes developments in the studies of dinosaurs in Italy particularly interesting is this country's geographical position. As it does today, the Mediterranean region at the time of the dinosaurs lay between the lands of the Northern Hemisphere and the African continent, in a strategic position for migratory movements and the colonization of new territories by terrestrial animals. Knowledge of the paleogeography of our planet has increased enormously thanks to studies carried out on dinosaurs. To give just one example, the hypothesis that India originated as a northward-drifting piece of Gondwana has been confirmed by the discovery of Indian fossils typical of the ancient southern supercontinent, such as abelisaurs (a group of Cretaceous theropods derived from ceratosaurs and found in South America, Africa, Madagascar, and India).

The footprints at Lavini di Marco show that several of the dinosaurs that trod the soil of paleo-Italy have Gondwanan affinities; that is, that they resemble the dinosaurs that lived in Africa. This suggests that during the Lower Jurassic, the carbonate platforms making up the western margin of the Tethys, which formed the "pieces" of what was later to become Italy, may have been partially above sea level and may have sporadically constituted a point of contact between the lands of Laurasia (to the north) and those of Gondwana (to the south). During the Middle and Upper Jurassic, Laurasia and Gondwana shifted farther apart, the "Italian" platforms sank, and the Tethys expanded westward. However, we know that the dinosaurs reappeared in this region during the Cretaceous, when the movement of the continents reversed direction and the Tethys began to close in on itself. The collision between the Eurasian Plate and the African Plate caused the Italian platforms to fuse and enabled different areas to emerge intermittently. It is quite likely that these movements led to the isolation of certain stretches of land, where, as was hypothesized for *Scipionyx,* new species of dinosaurs evolved. At other times, the Italian territory was probably used as a bridge linking its inhabitants with African (Apulia) or Eurasian (Carso) animal populations. These are other reasons why the study of Italian dinosaurs is arousing so much interest among paleontologists all over the world.

But where in Italy are other fossil discoveries most likely to come to light?

Many sedimentary formations of Mesozoic age, in which it was previously considered impossible to find remains of terrestrial vertebrates, have now revealed traces of the passage of the "terrible lizards," mainly in the form of footprints. Even though of marine origin, some portions of these formations bear signs of emersion, or in any case indications of the proximity of dry land. While these indications have not been totally ignored by geologists, their significance has undoubtedly been underestimated. Therefore, it could well be that many platform areas of the

Map of the most important dinosaur discoveries in Italy. As mentioned in the respective chapters, Besano and Cene are noteworthy for their marine reptiles and flying reptiles. *Graphics by F. Nodo.*

Italian Triassic, Jurassic, and Cretaceous Periods conceal dinosaur tracksites, and—who knows—perhaps not just tracks. As at Pietraroia and Saltrio, deposition of carcasses swept down from dry land may have occurred in other marine basins. In those of Besano, in the foothills of the Alps around Bergamo, and in Friuli, Triassic marine deposits continue to yield not only aquatic vertebrates but also the remains of terrestrial reptiles, and it is quite plausible that sooner or later a prosauropod or a primitive theropod will come to light, especially from the Carnian, Norian, and Rhaetian formations.

In Alpine and Apennine areas, there are marine carbonate rocks that contain "lenses" of sediment of a continental nature—deposits of a very limited distribution, of lacustrine or fluvial origin, many of which have received little or no paleontological investigation. More extensive deposits of continental Mesozoic rock outcrop in Sardinia, where nothing has been found, mainly because nobody has ever looked there for dinosaurs. In Apulia, the levels deposited at the emersion limit in the marine formations of the Cretaceous (i.e., the Altamura limestone) still look extremely promising due to their vast exposure. In the Altopiano delle Murge, these strata have sub-horizontal positions and faults that have caused only very slight dislocation. Consequently, they are characterized by considerable lateral continuity in no way comparable to that of the distorted alpine formations, which were literally shattered by orogenesis.

We are obviously more likely to be rewarded with new discoveries from localities where dinosaurs have already come to light, especially from Pietraroia, and from other Jurassic and Cretaceous lithographic limestones, which as yet have been studied very little, but which contain exceptionally well preserved fossils that are exposed here and there on the Apennines in the regions of Campania and Basilicata. Under the aegis of the Archaeological Superintendence of Salerno, plans are underway for further excavations by the University of Sannio and the Natural History Museum of Milan. "Ciro" couldn't have been alone, and while it is too fanciful to think of actually finding his mother, we can at least cherish the hope of finding the remains of an adult *Scipionyx*.

The possibility of discovering fossil eggs and nests is more remote. Based on the study of sediments in which such remains have been exposed in other parts of the world, it appears that dinosaurs preferred to nest farther inland, and certainly not where the tides could reach. The great majority of eggs found elsewhere were fossilized because they were covered by mudflows transported by the sudden flooding of rivers, which

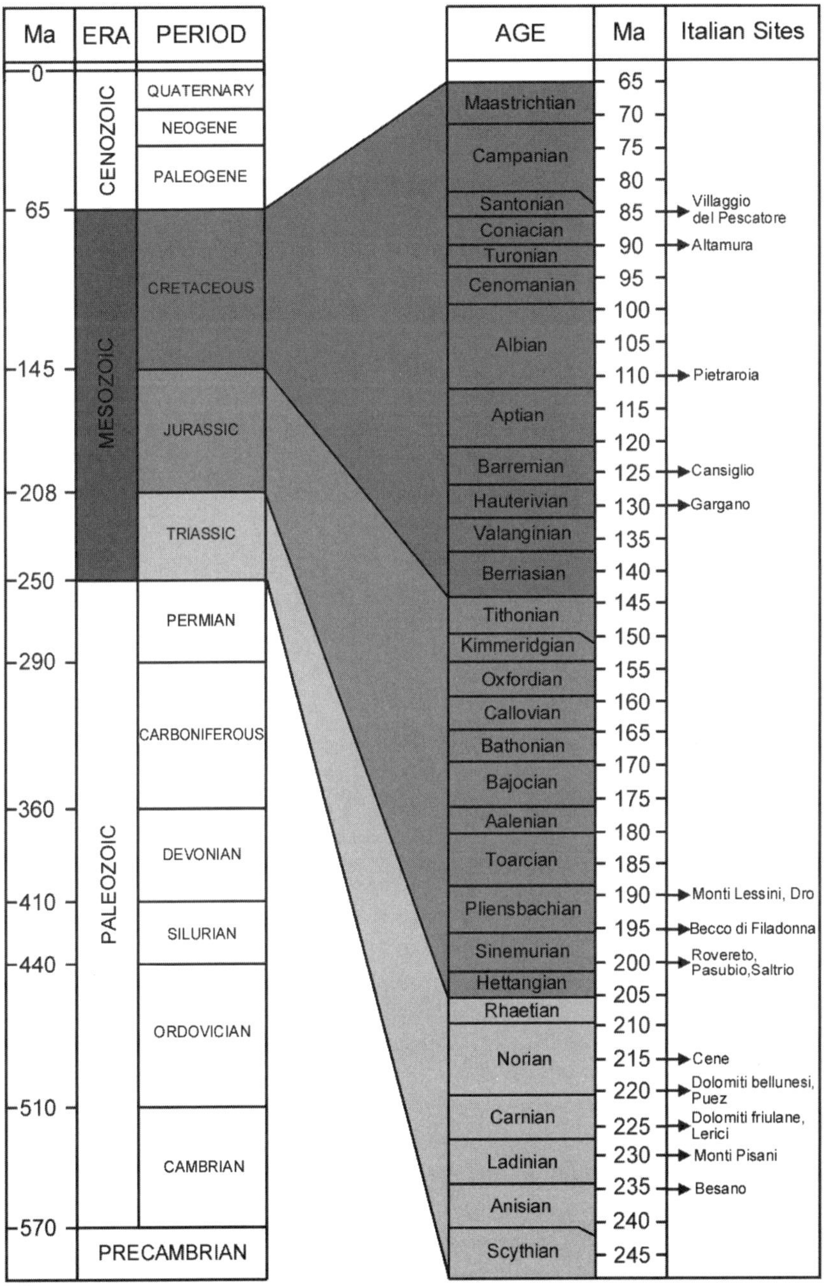


Geochronological table summarizing the distribution of the main fossil sites mentioned in the text. *Graphics by S. Maganuco and M. Yais.*

during the rainy seasons often broke their natural banks and flooded the plains inhabited by the dinosaurs. Mesozoic Italy was devoid of wide rivers because it had neither mountains nor volcanoes (the ones that were there formed islands with steep sides that did not slope down into plains). In fact, the lack of high hills or mountains prevented the accumulation of large quantities of rainwater, and consequently the formation of those wide, meandering rivers that could have transported enormous quantities of silt and clay onto the alluvial plains. Contrary to claims made by certain amateur paleontology enthusiasts, dinosaur eggs are very rarely found in marine deposits.[1]

Very likely, the new discoveries will lead to changes in the interpretation of Italy's geological history and the consequent re-plotting of the paleogeography of the eastern Tethys. The dinosaurs can't just have passed through Italy; they lived and reproduced there for thousands of generations. The "Bahamas-style" tropical islands were no longer enough; peninsulas, large islands, and extensive inland areas were needed, capable of supplying food for an enormous biomass such as the one represented by the hadrosaurs of the Triestine Karst, the theropod of Lombardy, and the ornithopods of Apulia.

Italians must start thinking seriously now about the possibility of finding dinosaurs in their own backyards. Discovery is always fortuitous, but luck needs a helping hand, which here results from gathering background information, looking around, and, when you are in doubt about whether you have come across something important, contacting the competent authorities. One thing for sure is that this book is only the start. The story of Italian dinosaurs has just begun.

# NOTES

## 1. Dinosaurs

1. Strictly speaking, time intervals are abstractions defined on the basis of successive layers of rocks. Thus we say that "Lower" Triassic rocks accumulated during the "Early" Triassic Period.

2. According to the modern conception of evolution, not all the dinosaurs disappeared. Some of them, many millions of years before the mass extinction that occurred at the end of the Cretaceous Period, gave rise to birds, which, from the phylogenetic point of view, can consequently be regarded as theropod dinosaurs.

3. An important consequence of this reasoning is that if two animals share certain novel characters, they can be grouped into the same category; the more such derived characters they have in common, the greater the probability that the two animals are related. Modern genealogical trees, referred to as cladograms, are diagrams that are constructed in accordance with the number of characters shared by the organisms under consideration.

4. Vertebrates are traditionally grouped into five classes, whose common family names are fishes, amphibians, reptiles, birds, and mammals. In actual fact, "fish" is an inaccurate term and is not considered valid by zoologists, in that it includes evolutionary stages of very different aquatic vertebrates: these include Agnatha (fishes devoid of jaws), such as the lampreys; Chondrichthyes (cartilaginous fishes), such as sharks and rays; and Osteichthyes (bony fishes), such as tunas, salmon, goldfish, and all the most evolved fishes. Birds, on the other hand, being dinosaurs with feathers, should be included among the reptiles, but their very distinct lifestyle has earned them a class of their own.

5. We have just mentioned that from the evolutionary viewpoint, birds are dinosaurs with feathers. So as not to create confusion, English-speaking authors have adopted the term "non-avian dinosaurs" to indicate "dinosaurs in the strict sense" each time they wish to exclude birds from the discussion.

6. In terms of structure, the famous *Eoraptor,* discovered in Argentina in the Ischigualasto Formation, is more primitive, but in terms of geological age it is slightly more recent.

7. For snakes, which have lost their legs, crawling has become the norm.

8. Modern reptiles are evocatively referred to as "cold-blooded." A more correct definition for the metabolism of lizards, snakes, crocodiles, and turtles is "ectothermy" (from the Greek *ektos,* outside, and *thermos,* heat). The body temperature of an ectotherm varies according to environmental conditions. Conversely, birds and mammals are "warm-blooded" animals, or rather endotherms (from the Greek *endothi,* inside), because they maintain a constant body temperature.

9. Binomial nomenclature was introduced by Carl von Linné, an eighteenth-century Swedish botanist better known as Linnaeus.

10. The term "carnosaurs" now includes only the Allosauroidea, a group of super-predatory theropods of the Jurassic and Cretaceous periods.

11. In 1994, Thomas Holtz Jr. of the University of Maryland revolutionized the conventional conception of coelurosaurs and bore out the intuition that had flashed through the brilliant mind of the German paleontologist Friedrich von Huene back in 1926. On comparing a hundred or so anatomical points, Holtz discovered that the family of tyrannosaurs has numerous features in common with the coelurosaurs, to the point that from an evolutionary standpoint, the tyrannosaurs can be considered as overgrown coelurosaurs, similar to the allosauroids, not by direct relationship but as a result of the previously mentioned phenomenon of parallel evolution called adaptive convergence.

12. A large animal heats up and cools down more slowly than a small one because temperature exchanges with the environment occur through the surface of the body. And since the surface grows by the square of the linear dimensions of a body but its volume grows by the cube, the larger the animal, the smaller the ratio between its volume and surface.

## 2. On the Tracks of the Dinosaurs

1. Sequences of tracks have been found in several areas of the United States and at Lavini di Marco, each composed of three parallel tracks, which are thought to have been produced by the claws of carnivorous dinosaurs that "scraped" against the bottom with the tips of their hind feet as they swam in shallow waters.

2. The need for a footprint to remain exposed to the air is the exact opposite of the condition required for fossilization of a carcass, which must be quickly covered with sediment. This is why, with very rare exceptions, no bones are found in footprint deposits and vice versa.

3. Very recently, new, better-preserved trackways of this type were recognized at the same site. Their pentadactyl U-shaped manual prints and tetradactyl narrow-gauged pedal prints, characterized by a long digit IV and a very short digit I, confirmed their affinity with the anatomy of the feet of *Vulcanodon*.

4. Gondwana is the name of the southern supercontinent which at that time joined South America, Africa, India, Australia, and Antarctica.

## 3. *Scipionyx*—"Ciro" to His Friends

1. The article with the formal description of the specimen, written by Cristiano Dal Sasso and Marco Signore, appeared in *Nature* 392, no. 6674 (March 26, 1998).

2. In reality, it is difficult to quantify the value of a scientific discovery in terms of money. Like all unearthed objects and works of art considered unique and inimitable, fossils such as *Scipionyx* are beyond price.

3. The archaic word "Pietraroja," probably of Spanish origin, which means "red stone," is still in use.

4. The limestone deposits of Pietraroia have been known since the nineteenth century for the beauty of the fossils they contain. The paleontological resources of the deposit were first mentioned in 1798 by Scipione Breislak in his book *Topografia fisica della Campania*. However, scientific studies began much later and focused immediately on the plentiful number of fossil fishes that came to light among the calcareous rocks. Between 1853 and 1864, Oronzo Gabriele Costa published the results of his studies in *Paleontologia del Regno di Napoli,* a series of volumes superbly illustrated with lithographic prints in the typical style of the period, in which he described the "Calcari ad Ittioliti di Pietraroia," a term that from then on was used as the official name of the geological formation. The deposit was subsequently studied by the geologists Geremia D'Erasmo (1915) and Bruno D'Argenio (1963), but in general with less interest than in the previous century. In more recent times, research was supervised by the Department of Paleontology of the University of Naples, which in 1982, in cooperation with the Museum of Natural Sciences of Turin, also conducted excavations in the locality of Le Cavere.

5. For experts in the field, it was tantamount to saying: "It's definitely a dinosaur." The presence of only three fingers with claws on the forelimbs is a characteristic unique to the more highly evolved theropod dinosaurs and the first birds. At that time, however, the only known specimen of the latter was *Archaeopteryx,* which came from an Upper Jurassic deposit in Germany.

6. Italian law provides that all archaeological and paleontological discoveries, including those found on private property, belong to the state. Consequently, it is prohibited, except for special concessions made to institutes for historical and scientific research, to look for and collect such specimens throughout Italian territory. The marketing and selling of Italian fossils is prohibited, but a recent ministerial provision allows them to be kept, on condition that the holder submit a self-certification declaring how many and what kinds of specimens he or she possesses and where they were found. The state may, however, appropriate the fossils held in private collections at any time and assign any specimens of scientific importance to public institutions. Although paleontology is a science very different from the historical-humanistic branches of learning (on the basis of which Italian archeologists are trained), according to Italian law the protection of fossils is the concern of the Archaeological Superintendences, each of which is authorized to appropriate the findings according to their territorial jurisdiction.

7. As mentioned, the operation was performed in Salerno but was entrusted to the Paleontology Laboratory of the Natural History Museum of Milan, at that time represented by Cristiano Dal Sasso and a freelance assistant, Sergio Rampinelli, who is highly skilled in the field of micro-preparations.

8. In 1996, *Microraptor* was still an unknown name. By pure chance, in 2002 it was independently re-created by other paleontologists to identify a small dromaeosaurid theropod unearthed in China.

9. The correct pronunciation is "shipiònyx," not "skipiònyx"; the English nickname "Skippy" is consequently misleading and has nothing to do with the origin of the name.

10. The actual title of the article, however, was decidedly more subdued: "Exceptional Soft-Tissue Preservation in a Theropod Dinosaur from Italy."

11. The impact on public opinion caused by the appearance of the dinosaur of Pietraroia in *Nature* was unprecedented for Italian paleontology. Interest in the specimen led to articles about its discovery being published on the front pages of at least 2,500 newspapers throughout the world.

12. Many bones are formed starting from two or more centers of ossification; in the embryo, these centers consist of very distinct nuclei surrounded by a cartilaginous mass, which converge and eventually fuse after birth as bone tissue gradually replaces the cartilaginous tissue. In immature vertebrates, the centers are separated by fine sutural lines between the bones, which are held together by the so-called conjugation cartilages: this is the point at which the bones continue to grow. If the young animal dies, the cartilage rapidly decomposes, often causing detachment of the centers of ossification; in *Scipionyx* this occurred in the tenth cervical vertebra, in the third to seventh dorsal vertebrae, in the fourth and fifth sacral vertebrae, and in the fifth and sixth caudal vertebrae.

13. Late ossification of the carpal bones in birds is understandable if we consider that the skeleton of the forelimbs is not necessary for survival in the early stages of life, because first the nestlings have to grow; only at a later date, practically in adulthood, do they learn to fly.

14. Although in a great majority of snakes the poison fangs are the ones nearest the front, embedded in the premaxillary bone, in the opisthoglyphs poison is injected by teeth located halfway along the upper jaw. One of the myths created by *Jurassic Park* is that dilophosaurs could "spit" poison. The idea was put forward by some paleontologist to explain the space between the premaxillary and the maxillary teeth, which exists not only in dilophosaurs but also in all the coelophysoids: it cannot be ruled out that this pocket housed a poisonous gland, but there are no data to prove this hypothesis either in the coelophysoids or in any other dinosaur.

15. In primates, and especially in humans, the bones from which the fingers branch off (metacarpals) are not fused together at the base; nor do they touch one another at the sides. Consequently, they can move in different planes (for example, try cupping the palm of your hand, as though you are holding water in it), giving the fingers greater freedom of movement (fold your thumb into the cup, covering it with your other fingers). Predatory dinosaurs, however, could extend their fingers and almost touch the back of the hand, which is much more than we can do.

16. In Fruili, emergence led to the production of paleosoils, residual clays, breccias, and karst phenomena; in the Matese area, the result was formation of sedimentary lacunae and vast deposits of bauxite, a rock rich in aluminum that can be formed only in a subaerial environment in hot, humid climates.

17. This is proved by the presence of deep-sea sediments, called turbidites, which were deposited during this period in areas that, due to their geographical position, separate the Molise-Apulian Platform limestone from the Latian and Campano-Lucanian Platform limestone.

18. In spite of this, in *Scipionyx* there is no trace of scales, feathers, or other tegumentary structures. This strange method of preservation depends upon the particular chemical conditions of the lagoon in which the dinosaur was fossilized, and will be examined further on.

19. Unfortunately, interpretation of the iron nodule as a fossil heart is not uni-

versally accepted. X-ray analysis carried out on the specimen shows that it had four cavities, corresponding to the two atria and the two ventricles of the heart of mammals and birds (reptiles have only one ventricle; crocodiles have two, but the septum that divides them is not complete). Should the news be confirmed, "Willo's" heart would be conclusive proof that at least some dinosaurs were warm-blooded animals (endotherms). In fact, what use would a heart with two atria and two ventricles—with complete separation between venous blood (non-oxygenated) and arterial blood (oxygenated)—have been if the animal did not need to burn oxygen continuously?

20. *New York Times,* March 26, 1998, p. A24

21. Although carnivorous dinosaurs have pneumatized bones—they were provided with hollow diverticula similar to those of birds—to date it has not been possible to demonstrate the existence of actual air sacs. The only case in which something appears to be seen is that of the incomplete skeleton of a coelurosaur of an indefinite species found well-preserved inside a nodular concretion extracted from the famous Brazilian fossil site of Chapada do Araripe. David Martill of the University of Portsmouth (England) believes that the cavities in the sediment are prints left by abdominal air sacs in front of the pubic bones.

22. In general, the intestines of carnivorous animals tend to be shorter and less complex than those of herbivores, which, since they need to digest cellulose, must be longer and more specialized. Consequently, according to some paleontologists, a short, thick digestive tract alone is not enough to indicate a high metabolism, and the question could be more complex.

23. Homologue of the clavicles of reptiles, the furcula is typical of birds and is well-known as the "wishbone"—the Y-shaped bone we find in a chicken breast. At the time of publication in *Nature,* the furcula had been found in very few dinosaurs, and among the coelurosaurs it was known only in the troodontids, in the oviraptorosaurs, and in *Archaeopteryx.*

24. The crystals are so small that it is not possible to classify their geometric form. Going by the color alone and in the absence of chemical analysis, it could also be another iron mineral, such as goethite, of which "Willo's" fossil heart proved to be composed.

25. The fossils of vertebrates are composed of crystals of apatite (a phosphate deriving from the transformation of bioapatite, of which the bones of the living animal are composed) and calcium carbonate (aragonite, calcite); more infrequently they are composed of silicates and metallic minerals.

26. Index fossils are so called because they are useful in indicating the age of the rocks in which they are contained. They are species that have evolved rapidly (and have consequently changed form many times, each form characterizing an interval of time) and have a wide geographic distribution. They are for the most part marine fossils, such as foraminiferans and ammonites. Since they are found in localities at a great distance from one another, index fossils are particularly useful for stratigraphic correlation—ascertaining whether the rocks that outcrop in two or more distant places were formed at the same time.

27. In *Scipionyx*'s thorax there is a small cluster of rounded bones that look like minute vertebrae. It is quite probable that the young *Scipionyx* fed on small lizards. A whole sphenodontid was found in the stomach of a *Compsognathus,* a coelurosaur

just a little larger than *Scipionyx*. To be able to capture lizards, these dinosaurs had to be really quick and agile.

28. Rudists were bivalve mollusks with heavy shells that built huge colonies in the seas of the Cretaceous Period, similar to present-day coral reefs.

29. Under exceptional circumstances, initial mineralization can take place within the space of a few days, and this is the only way that soft tissues can be preserved. However, as a rule, fossilization is a slow transformation requiring thousands of years. The speed of fossilization is not constant because it depends upon environmental conditions. In modern carbonate platforms, for example, cementation of sediments can take place within the space of a few weeks, with the result that we often see Coca-Cola cans on beaches that are not yet rusty but are already welded to blocks of sand.

## 4. "Antonio" and the Hadrosaurs of the Triestine Karst

1. Other paleontologists are not of the same opinion. Comparing the pelvic bones of the babies of *Maiasaura* with those of extant birds, it was found that the former were as well developed as those of precocial bird hatchlings. Therefore it is also possible that the dinosaur chicks were able to leave the nest, feed themselves, and return to the nest without receiving real care from their parents.

2. Natural mummification can also occur by immersion in very saline water, where the considerable difference in solute density causes the flow of fluids from the body by osmosis.

## 5. Jurassic, Italy: The Saltriosaur

1. Consequently, it is highly likely that the remains of the skeleton now lie ground up under the asphalt of some expressway in the Lombardy or Ticino area.

2. In March 2000, the remains of an even bigger theropod came to light in Patagonia. Another sensational fact is that the skeleton was discovered close to five others, three of which belonged to young individuals. Large carnivorous dinosaurs are always imagined as solitary animals, but the new discovery seems to point to a family behavior pattern similar to that of lions. But the record for the largest carnivorous dinosaur and terrestrial predator of all times seems to be held by the spinosaur, of which a 1-meter-long snout, housed at the Natural History Museum in Milan, has just recently been described. Taking proportions into consideration, this Moroccan fossil dating back to the Cretaceous belonged to an individual measuring 16 to 18 meters in length and weighing at least 8 tons.

3. The determination of this bone, in the past often mistaken for one of those thin little bones of the abdomen called gastralia and very well preserved in the little *Scipionyx,* is certain in this case: unlike the gastralia, the furcula of the Saltrio dinosaur is composed of two short "arms" devoid of longitudinal grooves, joined into a wide-open V to form an obtuse angle. In addition, at the top of the V the furcula is provided with a characteristic spine-like process, called a *hypocleidium,* while at the two ends it has a flat tip that acts as an articulation with two bones of the pectoral

girdle called coracoids. The furcula derives from the fusion of the two clavicles and occupies the same position.

## 6. The Besanosaur and Marine Reptiles

1. Even though cases of cannibalism have been observed among the ichthyosaurs, in this case it is difficult to imagine that a specimen with an extremely fine rostrum and very small teeth was able to swallow prey the size of an adult mixosaur (as indicated by the size of the vertebral discs contained in the stomach).

2. The scientific description of the new ichthyosaur, entitled "*Besanosaurus leptorhynchus* n. gen. n. sp., a New Shastasaurid Ichthyosaur from the Middle Triassic of Besano (N. Italy)" (in English), was published by Cristiano Dal Sasso and Giovanni Pinna in 1996 in the specialized journal *Paleontologia Lombarda* n.s. 4 (1996), ed. Società Italiana di Scienze Naturali & Museo Civico di Storia Naturale di Milano.

3. The discovery in the Canadian Rocky Mountains of a Triassic ichthyosaur measuring 23 meters in length was announced in 1999. It took six months for the skeleton, still trapped in the rock, to be collected from the field. It lacked only its hind flippers.

## 7. Flying Reptiles in Italy

1. As mentioned in Chapter 1, and as will be apparent in this chapter, pterosaurs are not the ancestors of birds, which, on the contrary, derive from small carnivorous dinosaurs.

2. Possibilities include the presence of reefs of calcareous algae that acted as a barrier to the open exchange of waters with the surrounding shallow seas of the carbonated platform, high temperatures, and eutrophication.

3. The scientific description of the first Italian pterosaur, entitled "*Eudimorphodon ranzii* gen. nov., sp. nov., uno pterosauro triassico," was published by Rocco Zambelli in 1973 in the *Rendiconti dell'Istituto Lombardo di Scienze e Lettere,* Classe B, vol. 107.

## 8. Gubbio

1. An exception that seems to represent a bizarre evolutionary experiment might be the Chinese *Microraptor.* Recently found in the Sihetun area, it shows long feathers also on the rear limbs, and at first glance was interpreted as a dinosaur that glided among the trees with "four wings."

## 9. The Future of Italian Dinosaurs

1. It is not unusual for some city dwellers on their way back from an excursion in the mountains to fill the trunks of their automobiles with dozens of spherical or

bizarrely shaped rocks to give to their friends or show to a paleontologist as dinosaur eggs or bones. We are always sorry to have to disappoint enthusiastic people, but however "perfect" the outer shape of these structures may be, their opaque-vitreous internal appearance shows them to be flint nodules, sedimentary concretions formed by isolation of silica inside limestone rocks (silicates and carbonates behave somewhat like oil and vinegar: they do not mix).

# SUGGESTED READING


Alexander, R. M. 1989. *Dynamics of Dinosaurs and Other Extinct Giants.* New York: Columbia University Press.

Alvarez, W. 1997. *T. Rex and the Crater of Doom.* Princeton, N.J.: Princeton University Press.

Bakker, R. T. 1986. *The Dinosaur Heresies: New Theories Unlocking the Mystery of Dinosaurs and Their Extinction.* New York: William Morrow & Co.

Benton, M. J., M. A. Shishkin, D. A. Unwin, and E. N. Kurochkin, eds. 2000. *The Age of Dinosaurs in Russia and Mongolia.* Cambridge: Cambridge University Press.

Bonaparte, J. F. 1998. *Los dinosaurios de la Patagonia argentina.* Buenos Aires: Museo Argentino de Ciencias Naturales.

Buffetaut, E. 1994. *Les dinosaures.* Paris: Presses Universitaires de France.

———, M. Delfino, and G. Pinna. 2001. "The Crocodilians, Pterosaurs and Dinosaurs from the Campanian-Santonian of Villaggio del Pescatore (Northeastern Italy): A Preliminary Report." Paper presented at the 6th European Workshop on Vertebrate Paleontology, Florence, September 19–22 2001, Università degli Studi di Firenze. Abstract volume.

Camanni, S., G. Rinaldi, and P. Spirito. 1994. *Dinosauri sulle Alpi.* Le guide di Alp. Torino: Vivalda Editori.

Carpenter, K. 1999. *Eggs, Nests, and Baby Dinosaurs: A Look at Dinosaur Reproduction.* Bloomington: Indiana University Press.

———, and P. J. Currie, eds. 1990. *Dinosaur Systematics.* Cambridge: Cambridge University Press.

———, K. F. Hirsch, and J. R. Horner. 1994. *Dinosaur Eggs and Babies.* New York: Cambridge University Press.

Carroll, R. L. 1988. *Vertebrate Paleontology and Evolution.* New York: John Wiley & Sons.

Chatterjee, S. 1997. *The Rise of Birds: 225 Million Years of Evolution.* Baltimore: Johns Hopkins University Press.

Chiappe, L. M., and L. M. Witmer, eds. 2002. *Mesozoic Birds: Above the Heads of Dinosaurs.* Berkeley: University of California Press.

Cita, M. B., and I. Premoli Silva. 1974. "Il limite Cretaceo-Paleocene e l'evoluzione della Tetide." *Rivista Italiana di Paleontologia e Stratigrafia,* Milano, Memoria XIV: 193–250.

Cornalia, E. 1854. "Notizie geologiche sul Pachypleura edwardsii." *Giorn. R. Ist. Lomb.,* Milano 6.

Costa, O. G. 1854–1856. *Paleontologia del Regno di Napoli.* Parte II. *Stabil.* Napoli: Tipogr. G. Cataneo.

Cox, B., B. Gardiner, R. J. G. Savage, and D. Dixon. 1988. *Macmillan Illustrated Encyclopedia of Dinosaurs and Prehistoric Animals: A Visual Who's Who of Prehistoric Life.* New York: Macmillan.

Currie, P. J., and E. Koppelhus. 1996. *101 Questions about Dinosaurs.* Mineola, N.Y.: Dover Publications.

———, and K. Padian, eds. 1997. *Encyclopedia of Dinosaurs.* New York: Academic Press.

Czerkas, S. J., and E. C. Olsen, eds. 1987. *Dinosaurs Past and Present.* Vols. I and II. Seattle: University of Washington Press.

Dal Sasso, C. 1995. *Animals: Origin and Evolution.* Austin, Tex.: Raintree Steck-Vaughn.

———. 2003. "Les dinosaures d'Italie." In *Les Dinosaures d'Europe.* Comptes Rendus Palevol, Academie des Sciences de Paris, II Série (Sciences de la Terre et Sciences Planetaires), II (1), Numéro Thématique. Paris: Elsevier.

———, and G. Pinna. 1996. "Besanosaurus leptorhynchus n. gen. n. sp., a New Shastasaurid Ichthyosaur from the Middle Triassic of Besano (N. Italy)." *Paleontologia Lombarda,* n.s. 4, ed. Società Italiana di Scienze Naturali & Museo Civico di Storia Naturale di Milano.

———, and M. Signore. 1998. "Exceptional Soft-Tissue Preservation in a Theropod Dinosaur from Italy." *Nature* 392: 383–387.

Dalla Vecchia, F. M. 2001. "The Mesozoic Periadriatic Carbonate Platforms as Terrestrial Ecosystems: The Vertebrate Evidence." Paper presented at the Seventh International Symposium on Mesozoic Terrestrial Ecosystems, Buenos Aires, September 26–October 1, 1999. Abstracts.

Desmond, A. J. 1976. *The Hot-Blooded Dinosaurs.* New York: The Dial Press.

Dong, Z. 1988. *Dinosaurs from China.* Beijing: China Ocean Press.

Farlow, J. O., and M. K. Brett-Surman, eds. 1997. *The Complete Dinosaur.* Bloomington: Indiana University Press.

Fastovsky, D. E., and D. B. Weishampel. 1996. *The Evolution and Extinction of the Dinosaurs.* Cambridge: Cambridge University Press.

Gans, C., ed. 1969. *Biology of the Reptilia.* Vol. 1: *Morphology A.* New York: Academic Press.

Gans, C., and K. A. Gans, eds. 1978. *Biology of the Reptilia.* Vol. 8: *Physiology B.* New York: Academic Press.

Grassé, P. P., ed. 1970. *Traité de Zoologie.* Tome XIV: *Reptiles: Charactères generaux et anatomie.* Paris: Masson.

Gauthier, J., and L. F. Gall, eds. 2001. *New Perspectives on the Origin and Early Evolution of Birds.* Proceedings of the International Symposium in Honor of John H. Ostrom. New Haven, Conn.: Peabody Museum of Natural History.

Hecht, M. K., J. H. Ostrom, G. Viohl, and P. Wellnhofer, eds. 1985. *The Beginning of Birds*. Proceedings of the International Archaeopteryx Conference, Eichstätt, 1984. Eichstätt: Freunde des Jura-Museum Eichstätt.

Hildebrand, M. 1988. *Analysis of Vertebrate Structure*. 3rd ed. New York: John Wiley & Sons.

Holtz, T. R. Jr. 2000. "A New Phylogeny of the Carnivorous Dinosaurs." In B. Perez-Moreno, T. Holtz, J. L. Sanz, and J. J. Moratalla, eds., *Aspects of Theropod Paleobiology*, 5–61. Gaia, no. 15. Lisbon, Portugal: Museu Nacional de História Natural, Universidade de Lisboa.

Horner, J. R., and D. Lessem. 1990. *The Complete T. rex*. London: Souvenir Press.

Kardong, K. V. 1998. *Vertebrates: Comparative Anatomy, Function, Evolution*. New York: McGraw-Hill.

Lambert, D. 1990. *Dinosaur Data Book*. New York: Avon Books.

Leiggi, P., and P. May, eds. 1994. *Vertebrate Palaeontological Techniques*. Vol. 1. Cambridge: Cambridge University Press.

Leonardi, G., and P. Mietto, eds. 2001. *Dinosauri in Italia*. Pisa: Accademia Editorale.

Ligabue, G., ed. 1994. *Il Tempo dei dinosauri*. Milano: Quaderni de Le Scienze.

Lockley, M. 1994. *Tracking Dinosaurs: A New Look at an Ancient World*. New York: Cambridge University Press.

———, and C. Meyer. 2000. *Dinosaur Tracks and Other Fossil Footprints of Europe*. New York: Columbia University Press.

Lucas, S. G. 1994. *Dinosaurs: The Textbook*. Dubuque, Iowa: William C. Brown Publishers.

Lyman, R. L. 1994. *Vertebrate Taphonomy*. Cambridge: Cambridge University Press.

Marshall, A. J., ed. 1960–1961. *Biology and Comparative Physiology of Birds*. Vol. I. New York: Academic Press.

McGowan, C. 1992. *Dinosaurs, Spitfires and Sea Dragons*. Cambridge, Mass.: Harvard University Press.

Nicosia U., M. Marino, N. Mariotti, C. Muraro, S. Panigutti, F. M. Petti, and E. Sacchi. 1999. "The Late Cretaceous Dinosaur Tracksite near Altamura (Bari, Southern Italy)." *Geologica Romana* 35: 237–247.

Norman, D. B. 1985. *The Illustrated Encyclopedia of Dinosaurs*. London: Salamander Books.

Novacek, M. 1996. *Dinosaurs of the Flaming Cliffs*. New York: Anchor Books.

Novas, F. E., and R. E. Molnar, eds. 1996. *Proceedings of the Gondwanan Dinosaur Symposium*. Brisbane: Queensland Museum.

Ostrom, J. H. 1969. "Osteology of Deinonychus antirrhopus: An Unusual Theropod from the Lower Cretaceous of Montana." *Bulletin of the Peabody Museum of Natural History* 30: 1–165.

Padian, K., and D. Chure, eds. 1989. *The Age of Dinosaurs.* Short Courses in Paleontology, no. 2. Knoxville, Tenn.: Paleontological Society.

Paul, G. S. 1988. *Predatory Dinosaurs of the World.* New York: Simon and Schuster.

———. 2002. *Dinosaurs of the Air.* Baltimore: Johns Hopkins University Press.

———, ed. 2000. *The Scientific American Book of Dinosaurs.* New York: St. Martin's Press.

Pinna, G., and G. Teruzzi. 1991. "Il giacimento paleontologico di Besano." *Natura* 82 (1).

Preiss, B., and R. Silverberg, eds. 1992. *The Ultimate Dinosaur: Past—Present—Future.* New York: Bantam Books.

Romer, A. S. 1954. *Osteology of the Reptiles.* Chicago: University of Chicago Press.

Rosenberg, G. D., and D. L. Wolberg, eds. 1994. *Dino Fest.* The Paleontological Society Special Publication, no. 7. Indianapolis: Geology Department, Indiana University–Purdue University, and [Knoxville]: Paleontological Society.

Starck, J. M., and R. E. Rucklefs. 1998. *Avian Growth and Development.* Oxford: Oxford University Press.

Stewart, W. N., and G. W. Rothwell. 1993. *Paleobotany and the Evolution of Plants.* 2nd ed. Cambridge: Cambridge University Press.

Tanke, D. H., and K. Carpenter, eds. 2001. *Mesozoic Vertebrate Life.* Bloomington: Indiana University Press.

Taylor, T. N., and E. L. Taylor. 1993. *The Biology and Evolution of Fossil Plants.* Englewood Cliffs, N.J.: Prentice Hall.

Thomas, R. D. K., and E. C. Olsen, eds. 1980. *A Cold Look at Warm-Blooded Dinosaurs.* Boulder, Colo.: Westview.

Thomason, J., ed. 1995. *Functional Morphology in Vertebrate Paleontology.* Cambridge: Cambridge University Press.

Walker, F. W. Jr., and K. F. Liem. 1994. *Functional Anatomy of the Vertebrates: An Evolutionary Perspective.* Philadelphia: Saunders College Publishing.

Weigelt, J. 1989. *Recent Vertebrate Carcasses and Their Paleobiological Implications.* Chicago: University of Chicago Press.

Weishampel, D. B., P. Dodson, and H. Osmolska, eds. 1990. *The Dinosauria.* Berkeley: University of California Press.

Wellnhofer, P. 1991. *The Illustrated Encyclopedia of Pterosaurs.* London: Salamander Books.

Wild, R. 1978. "Die Flugsaurier (Reptilia, Pterosauria) aus der Oberen Trias von Cene bei Bergamo, Italien." *Boll. Soc. Paleont. It.*, Modena 17, no. 2: 176–256.

Wilford, J. N. 1986. *The Riddle of Dinosaurs.* New York: Knopf.

Zambelli, R. 1973. "*Eudimorphodon ranzii* gen. nov., sp. nov., uno pterosauro triassico." *Rendiconti dell'Istituto Lombardo di Scienze e Lettere,* Classe B, 107: 27–32.

Zappaterra, E. 1990. "Carbonate Paleogeographic Sequences of the Periadriatic Region." *Boll. Soc. Geol. Ital.* 109: 5–20.

**CRISTIANO DAL SASSO** is Head of the Laboratory of Paleontology at the Museum of Natural History in Milan and technical coordinator of the excavations at the Triassic fossil site of Besano. He has described three new kinds of reptiles (*Aphanizocnemus, Besanosaurus,* and *Scipionyx*) and has authored numerous popular reports and scientific papers, which have been published in international reviews such as *Nature, Science,* and the *Journal of Vertebrate Paleontology.*

**GIUSEPPE BRILLANTE** is a freelance journalist. He writes for various scientific reviews, including *Airone* and *Newton.* He lives and works in Milan.